CRC

PHYSICAL AND ENGINEERING ASPECTS
OF THERMAL POLLUTION

authors:

Frank L. Parker
Peter A. Krenkel
Vanderbilt University
Nashville, Tennessee

published by:

A DIVISION OF
THE CHEMICAL RUBBER CO.
18901 Cranwood Parkway • Cleveland, Ohio 44128

CRC MONOSCIENCE SERIES

The primary objective of the CRC Mono-science Series is to provide reference works, each of which represents an authoritative and comprehensive summary of the "state-of-the-art" of a single well-defined scientific subject.

Among the criteria utilized for the selection of the subject are: (1) timeliness; (2) significant recent work within the area of the subject; and (3) recognized need of the scientific community for a critical synthesis and summary of the "state-of-the-art."

The value and authenticity of the contents are assured by utilizing the following carefully structured procedure to produce the final manuscript:

1. The topic is selected and defined by an editor and advisory board, each of whom is a recognized expert in the discipline.

2. The author, appointed by the editor, is an outstanding authority on the particular topic which is the subject of the publication.

3. The author, utilizing his expertise within the specialized field, selects for critical review the most significant papers of recent publication and provides a synthesis and summary of the "state-of-the-art."

4. The author's manuscript is critically reviewed by a referee who is acknowledged to be equal in expertise in the specialty which is the subject of the work.

5. The editor is charged with the responsibility for final review and approval of the manuscript.

In establishing this new CRC Monoscience Series, CRC has the additional objective of attacking the high cost of publishing in general, and scientific publishing in particular. By confining the contents of each book to an *in-depth treatment* of a relatively narrow and well-defined subject, the physical size of the book, itself, permits a pricing policy substantially below current levels for scientific publishing.

Although well-known as a publisher, CRC now prefers to identify its function in this area as the management and distribution of scientific information, utilizing a variety of formats and media ranging from the conventional printed page to computerized data bases. Within the scope of this framework, the CRC Monoscience Series represents a significant element in the total CRC scientific information service.

B. J. Starkoff, President
THE CHEMICAL RUBBER Co.

This book originally appeared as part of an article in *CRC Critical Reviews in Environmental Control,* a quarterly journal published by The Chemical Rubber Co. We would like to acknowledge the editorial assistance received by the Journal's co-editors, Professor Richard G. Bond and Dr. Conrad P. Straub, both at the University of Minnesota. Mr. Donald P. Stevens served as referee for this article.

INTRODUCTION

Thermal pollution problems will intensify as electricity usage increases unless substantial changes are made in the mode of discharge of the heated condenser cooling waters. By far, the major source of heated waters is from central electricity generating stations. Not only will electricity usage increase but, in addition, the problem will grow more than linearly due to the introduction of larger sizes of individual generating units and plant sizes and due to the trend to nuclear power.

Nuclear power plants discharge as much as 50% more hot water to a stream as do equivalent fossil fuel power plants. Possibilities of higher efficiencies from advanced technology for central power stations seem remote. Fuel cells, magnetohydrodynamics, electrogasdynamics and thermonuclear fusion and weather modification are not yet ready for routine central station use.

The effects of the increased amounts of heated water discharge to the streams are manifold; possibly, most obvious is the induction of stratified flow conditions and the lowering of the capacity of water to hold oxygen, increased reaeration rates, and higher metabolic rates. The net result is that the oxygen balance due to reaeration and deaeration is smaller. The addition of heat to a stream or reservoir is equivalent to reducing the waste assimilative capacity of that stream. With the earlier exhaustion of oxygen, reducing conditions may occur in the reservoir waters and cause iron and manganese compounds to enter into solution. At higher temperatures, chemical reactions proceed at a faster pace; and, therefore, minor savings in the cost of treating water are possible. Waste heat can be beneficially used to heat water from the hypolimnion to the proper temperature for irrigation. Waste heat could also be used to keep some of our icebound rivers open throughout the entire year. Increased yields of seafoods might be possible by warming their waters.

In general, most methods of preventing excessive heating of our waterways have depended upon cooling ponds and cooling towers. If the recently announced limit of one degree Fahrenheit rise in Lake Michigan holds nationally, then all power plants will have to go to recirculating or air coolant systems. Cooling ponds will continue to be used where land values are low, but the major emphasis will be upon cooling towers. Hyperbolic natural draft towers have begun to be built in larger numbers in the United States in recent years. Though no air cooled condensers for central electricity generating stations have yet been built in the U.S. in water short areas, they may soon appear. The cost of cooling all the condenser cooling waters of central electric generating stations by the year 2000 by wet induced draft cooling towers will be 11 billion dollars. The benefits to be derived from relief of thermal pollution have not yet been quantified.

THE AUTHORS

Peter A. Krenkel is Professor and Chairman of the Department of Environmental and Water Resources Engineering at Vanderbilt University. He has been at Vanderbilt since 1960.

Dr. Krenkel is an active consultant to industry and to many government agencies in the United States and abroad, advising on various environmental engineering problems, including water-quality management, distribution systems, drainage problems, and disposal systems. He was awarded the Rudolph Hering Medal for Most Valuable Contribution to Research in Sanitary Engineering in 1963.

Dr. Krenkel's degrees are from the University of California at Berkeley (B.S. in C.E., 1956; M.S., 1958, and Ph.D., 1960, in Civil and Sanitary Engineering). He is a member of the American Society of Civil Engineers, Water Pollution Control Federation, National Society for Professional Engineers, Air Pollution Control Association, U.S. Public Health Service Reserve, American Water Works Association, American Association for Professors in Sanitary Engineering, and American Institute of Chemical Engineers. He is editor with Frank L. Parker of *Biological Aspects of Thermal Pollution*, Vanderbilt University Press, 1969.

Frank L. Parker is Professor of Environmental and Water Resources Engineering at Vanderbilt University.

Before joining the Vanderbilt faculty in 1967, Dr. Parker was Chief of the Radioactive Waste Disposal Research Section at the Oak Ridge National Laboratory and, simultaneously, was engaged as Co-ordinator of the Clinch River Study, an ecological survey involving several agencies of the federal government and of the State of Tennessee. He has also had wide experience as a consultant to the United States government and many foreign governments on waste disposal problems.

Dr. Parker attended the Massachusetts Institute of Technology (B.S. in C.E., 1948) and Harvard University (M.S. in C.E., 1950; Ph.D. in Water Resources, 1955). He is a member of the American Society of Civil Engineers, Health Physics Society, American Association for the Advancement of Science, Water Pollution Control Federation, and American Geophysical Union. He is editor with Peter A. Krenkel of *Engineering Aspects of Thermal Pollution*, Vanderbilt University Press, 1969.

TABLE OF CONTENTS

INTRODUCTION

Thermal pollution, thermal enhancement, or thermal enrichment: What is the proper description of the waste heat from cooling condensers? To answer this question, we need to know the quantity of excess heat available, the effect of this heat, and alternative means of using or disposing of this heat. The cooling water required in 1965 for our central generating stations for even a 20°F rise in our rivers is greater than the minimum daily flow of the Mississippi River at Vicksburg.[1] We can note that any large scale modification of the thermal regime of rivers and reservoirs will change its ecology and usually for the worse. The most obvious and destructive change is the heat killing of fish. Less obvious but possibly equally destructive in the long run may be the sublethal effects. Though only a small number of fish kills due to thermal changes have occurred to date, if unrestricted use of streams and reservoirs for cooling purposes were allowed, the number would greatly increase. In addition, discharges of heated waters are equivalent in their effect to the discharges of organic wastes in lowering the assimilative capacity of a stream. Therefore, our concern with this excess heat is primarily to prevent a potential major pollution problem from occurring.

The greatest single source of man-made heat addition to our waterways is from electric generating plants. Analysis of steam electric cooling discharges indicate an average 15°F rise in water temperatures after passing through condensers.[2] The amount of water withdrawn for this purpose is approximately 40 trillion gallons per year which is roughly 10% of the total flow of waters in U.S. rivers and streams per year.[3] Industrial cooling water discharges are approximately 10 trillion gallons per year.[4] The magnitude of the industrial cooling water volume, however, is fraught with a great deal of uncertainty because of the great difficulty in evaluating and obtaining definitive information on industrial cooling water uses. The plants differ widely among themselves, even within the same industry. The temperature of the discharge waters is less critical and consequently the information is not as satisfactory as that available from the steam electric generating plants. We can, however, conclude that this source will not be as concentrated nor as troublesome as discharges from electric generating plants. Their relative impact can be assessed by the Federal Water Pollution Control Administration estimated costs of total cash outlays from 1969 through 1973 for cooling discharges from steam generating stations of approximately 2 billion dollars, and for cooling manufacturing discharges, 700 million dollars.[5]

In addition to the power plant and industrial waste heat additions to waterways, modest increases in the heat load of the waters are caused by municipal and irrigation return waters. Published information on the temperature rise of water from intake plants through the discharge point of the sewage treatment plant is sparse. On the Thames Estuary it was reported that sewage effluents are responsible for 9% of the heat additions to the stream.[6] Power station discharges are responsible for 74.5%, industrial effluents 6%, freshwater discharges 6.5%, and biochemical activity 4% of the total heat rise. For Nashville, Tennessee, for a typical year, the average temperature rise through the system is 3.5°F.[7]

Return flows from irrigation fields are also a thermal pollutant. As in any environmental situation the actual rise is a function of many things, including the initial water temperature, volume of abstraction, velocity of flow, length of time in field, and climatic conditions. For one site in the state of Washington, the rise has been found to be from 9° to 19°F.[8]

Central Electricity Generating Stations

Growth in Central Station Energy Requirements

The really significant heat loads result from the discharge of condenser cooling water from the steam electrical generating plants. Electrical power generation in the United States has doubled every ten years since 1945 and all indications are that the rate of increase will be even greater during the next few decades. Furthermore, the problems associated with these con-

centrated heat loads are compounded by the increased size of individual power plants and the greater quantities of heat discharged by plants of equivalent size when nuclear fueled reactors are used.

It has been estimated that 95% of the centrally generated electricity that is produced by fossil fuel today will be 65% nuclear by the year 1980.[9] As will be shown subsequently, large nuclear power plants currently require approximately 50% more cooling water for a given temperature rise than do fossil fuel plants of equal size. It is expected that improved technology will reduce this added requirement to 25% by the year 1980. Development of nuclear breeder reactors will reduce the amount of heat to be dissipated to that of fossil-fueled plants; however, these are not expected to be available until the 1980's.[10]

The Atomic Energy Commission (AEC)[11] stated that as of June, 1968, 13 nuclear central-station electric power reactors were in operation, 31 were being built, and 40 were being planned. The increase in size of the newer power stations is indicated by noting that the nuclear power plants planned for operation by 1973 average 624 megawatts per unit while fossil fueled plants retired between 1962 and 1965 averaged 22 megawatts per unit.[12]

Figure 1 (see page 11) shows the anticipated growth of the power industry until the year 2020.[13] One should note the relatively small percentage of power production that is and will be contributed by sources other than steam generation.

In 1964, hydroelectric plants provided 19% of the capacity and energy produced and indications are that this percentage will decline as shown in Figure 1. Though economical sites for electric power development are being steadily eliminated, potential sites for pumped storage installations still exist in the United States.[14] The optimal use of pumped storage is for peaking power and reserve capacity since the large steam electric plants being constructed operate best at high plant factors which are complemented by the low plant factors of the pumped storage plants.

The rate of growth in electric usage is due to the increase in population and the greater use of electricity per family unit as we go to electric

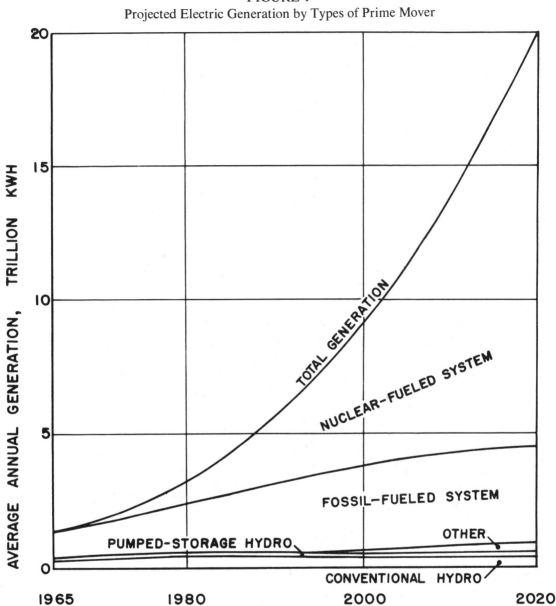

FIGURE 1
Projected Electric Generation by Types of Prime Mover

space heating, wider use of air conditioning, and a plethora of electric appliances of dubious utility. Another factor which intensifies the thermal pollution problem is that over 50% of the generating capacity ordered recently has been nuclear, primarily light water reactors. The thermal efficiencies of the light water reactors contracted for recently have been on the order of 34% in comparison to the 40% for equivalent fossil fueled plants. As long as light water reactors predominate, and this seems to be likely for a number of years, we can expect that the thermal efficiencies of the nuclear power plants will be below that of equivalent fossil fuel electric generating plants. The reason for this is the lower temperatures and pressures in reactor shells. This will continue because the AEC has dropped many of the projects for advanced converters and for super-heat and reheat of the light water fuel reactors because of metallurgical problems. The AEC seems to be going for fast breeder reactors as its next major step.

However, the preponderance of nuclear commitments may be on the wane as the rise in costs of nuclear power plants has been considerably above that for fossil fueled plants since the low of the Browns Ferry first and second units as shown in Table 1 (see below).[15] At that time nuclear power plants were competitive with fossil fuels at 16 cents per thousand BTU's. A recent large nuclear power plant, the Surry, is competitive with fossil fuel at 24.83 cents per thousand BTU's which shows the rate at which nuclear power costs have increased relative to fossil fueled plants.

Growth in Size

We can also expect that the thermal problems associated with cooling water discharges from steam electric generating plants will even further exceed the problems from industrial sources because it appears that industrial plants have by and large already reached the maximum economically efficient size and decentralization seems to be more important now, whereas, in steam electric generating plants, we may note that in the last fifteen years the maximum size of the generating units has risen from 208 Mw to 1100 Mw. In addition, the increase in size of individual units has made a great difference in the *intensity* of the thermal pollution problem. Although the total amount of heat disposed of to the environment may not increase when a single 1100 Mw plant is built rather than two 550 Mw plants at separated sites, the impact is far greater because the amount of dilution immediately available is greatly decreased and the heat loss to the atmosphere between sites does not occur. The units being installed at Brown's Ferry will be 1100 megawatts and consequently, even if the unit rate of release of heat were the same, the increase in size of units means that a greater amount of heat is being discharged at a point to the same size stream.

Efficiency—Heat Rate

It is also true, of course, that the efficiency of these units has increased over the years. The average net heat rate dropped from 25,000 BTU per kilowatt hour in 1925 to 10,453 BTU per kilowatt hour in 1965, so we have seen a net heat rate increase at a 2.8% annual rate over the last 40 years, whereas total power generation has increased at a 7.2% annual rate and the unit sizes have increased at an 18% annual

TABLE 1
NUCLEAR POWER COSTS[15]

Year	Atomic Unit	Size (megawatts)	Cost of Station (per kilowatt)	Total Production Cost (mills per kilowatt-hour)	Competitive Point (cents per million B.T.U.)
1964	Post-Oyster Creek	605.0	$139	4.48	26.3
1965	Post-Dresden 2	800.0	$123	4.42	24.0
1966	Browns Ferry 1, 2	1,100.0	$115	3.78	16.0
1967	Browns Ferry 3	1,100.0	$132	4.04	19.0
1967	Bridgman	1,100.0	$139.50	4.30	21.95
1967	Surry	815.5	$152.55	4.55	24.83

rate over the last 15 years. At present we seem to be approaching the upper limits of efficiency in fossil fuel steam generation at a heat rate of about 8500 BTU's/kwhr, which is an overall efficiency of about 40%.

Current generally accepted maximum operating conditions for conventional thermal stations are 1000°F and 3500 psi, with a corresponding heat rate of 8700 BTU per kwh, 3413 BTU's resulting in power production and 5287 BTU's being wasted. Plants have been designed for 1250°F and 5000 psi; however, metallurgical problems have held operating conditions to lower levels.

Nuclear plants operate at temperatures from 500 to 600°F and pressures up to 1000 psi, resulting in a heat rate of approximately 10,500 BTU/kwh. Thus, for nuclear plants, 3413 BTU's are for useful power production and 7087 BTU's are wasted.

Table 2 (see above) details the increasing efficiencies that have resulted from significant improvements in the conventional steam cycle.[16] Future heat rate gains will be minimal, however, because supercritical pressures of 3500 to 5000 psi and temperatures of 1100 to 1200°F are reflected in the data for 1961 and 1962. Present day central station average thermal efficiency is said to be 33% and the present best performance, 42%.

This is the most economic efficient rate. It is possible to arrive at a higher efficiency than this but the total costs are greater and, therefore, this is most likely the best plant heat rate that can be expected in the near future. If we compare this 40% efficiency with the Carnot cycle, using again the most economic temperature of 1000°F and the average temperature of surface waters in coterminous U.S. of 55°F, one then arrives at a Carnot cycle efficiency of 65%. Therefore, our 40% efficiency for the most economically efficient plant is actually 61% of the maximum possible efficiency. It cannot be expected that in this cycle the efficiencies can be greatly increased.

One should perhaps not compare the steam cycle with the Carnot cycle, since the Rankine cycle more nearly approximates the actual cycle in a steam plant. However, the Carnot cycle is the reversible cycle possessing the highest possible thermal efficiency. In the Carnot cycle heat

TABLE 2

NET HEAT RATES FOR STEAM-ELECTRIC
GENERATING STATIONS
1925 TO 1964[16]

Year	Best Plant[2] BTU/KWH	Best System BTU/KWH	U.S.[1] Average BTU/KWH
1925	15,000	[3]NA	25,000
1930	12,900	NA	19,800
1935	12,300	NA	17,850
1936	10,954	NA	17,800
1937	10,779	NA	17,850
1938	10,788	NA	17,450
1939	10,770	NA	16,700
1940	10,729	NA	16,400
1941	10,606	NA	16,550
1942	10,596	NA	16,100
1943	11,021	NA	16,000
1944	10,689	NA	15,850
1945	10,345	NA	15,800
1946	10,608	12,715	15,700
1947	10,600	NA	15,600
1948	10,588	NA	15,738
1949	10,437	NA	15,033
1950	9,378	11,876	14,030
1951	9,379	11,676	13,641
1952	9,303	11,665	13,361
1953	9,329	11,185	12,889
1954	9,113	10,660	12,180
1955	9,151	10,270	11,699
1956	9,106	9,780	11,456
1957	9,118	9,705	11,365
1958	9,130	9,760	11,085
1959	9,011	9,620	10,970
1960	8,975	9,590	10,760
1961	8,760	9,363	10,650
1962	8,588	9,390	10,558
1963			10,482
1964			10,462

[1]Exclusive of Alaska and Hawaii.
[2]Plants in service full year.
[3]NA - Not Available.

is supplied at the highest temperature, rejected at the lowest temperature, and undergoes adiabatic expansion and compression. The efficiency of the Carnot cycle may be computed by the differences in the absolute temperature of the heat added and rejected, divided by the absolute temperature of the heat added. Therefore, as indicated above, with the heat added at a temperature of (1000° + 460°) 1460°R and rejected at (55° + 460°) 515°R, the thermal efficiency is 65%.

The Rankine cycle is more nearly representative of a modern steam plant because it does not include an adiabatic compression step. Instead, heat is added to the feed water, bringing it to a boil, and converting it to steam. The distinguishing contrast is between the isentropic compression of the Carnot cycle and the heat added to the pressurized water until the critical point.

The best efficiencies for present day light water reactors are about 34%, which means that the overall amount of heat discharged to receiving waters from nuclear fuel power generating stations is greater than from fossil fuel power generating stations, but the problem itself is not substantially changed. We can, however, look forward to improved efficiencies in the high temperature gas reactors and in the breeder reactors when they do come on line. Rosenthal et al.[17] estimate that the high temperature gas cooled reactor of 1000 Mw will have a thermodynamic efficiency of 44.4%. Best estimates are that the net thermal efficiency of a 1000 Mw molten salt breeder reactor will be 44.9%.

From this fairly extensive discussion of the future growth of the thermal pollution problem and the sources of heat, we can see that there is not much hope with the present and foreseeable technology to decrease to any substantial degree the total amount of heat rejected to streams from these central steam generation plants. The interdepartmental energy study under the direction of Ali Cambel[18] stated, "Further cost reductions in conventional central power stations will be difficult to attain in view of the high internal efficiencies and the economies of plant size already realized."

Cooling Water Requirements

It has been estimated that cooling water requirements will increase from 50 trillion gallons per year in 1968 to 100 trillion gallons per year by 1980, and that this will be approximately one-fifth of the total runoff in the coterminous United States. Actually, the major portion of water may be used for other beneficial purposes or reuse. Examination of Table 3 (see below) which is the Water Resources Council's estimate of cooling water requirements and consumptive use based on a temperature rise of 15°F, demonstrates the consumptive use per kwhr in fossil-fueled steam electric power plants.[19] The predicted decrease in unit water requirements is based on improved technology.

Table 4 (see page 15)[20] gives the total daily fresh water requirements for cooling and consumptive use for steam electric power generation for the years 1965 to 2020.

Improving Efficiency of Central Station Steam Electricity Generation Stations

To increase the efficiencies of central steam electricity generating stations, it is possible to

TABLE 3

AVERAGE CONDENSER WATER REQUIREMENT AND CONSUMPTIVE
USE FOR FOSSIL-FUELED, STEAM-ELECTRIC POWER PLANTS,
1965 - 2020
Gallons per KWHR[19]

Year	Condenser Requirements	Consumptive Use		
		Once Through	Cooling Ponds	Cooling Towers
1965	40	0.3	0.4	0.5
1980	35	0.2	0.3	0.4
2000	30	0.15	0.25	0.35
2020	25	0.1	0.2	0.3

increase boiler pressure, decrease the back pressure, increase the temperature of the inlet steam, increase the temperature of the feed water by using a portion of the steam taken from turbine prior to the condenser, and reheat steam taken from the turbine and reinjected into the turbine after being reheated in the boiler. The approximate increases in efficiency for each of these changes is shown in Table 5 (see above).[21]

A schematic of a steam plant incorporating all of the efficiency increasing items is shown in Figure 2 (see page 16).[22]

The idealized temperature—entropy diagram is shown in Figure 3 (see page 17), where:
Boilerfeed

a - b Water is pressurized in pump.

b - c Heat is added to the feed water through regenerative feed water heaters.

c - d Heat is added in the boiler, turning water to steam.

d - 1 Heat added in boiler raising steam temperature to superheated steam.

1 - 2 Steam expands in high pressure turbine until steam starts to condense.

2 - a Steam condensed in condenser.

The area inside the heavy lines (a-b-c-d-1-2-a) is the work done, whereas the area under a-b-c-d-1-2 is the heat added. It can readily be seen that if the area under a - 2 (heat rejected) is held constant, then any increase in the area above will increase the thermodynamic efficiency though not necessarily the economic efficiency.

TABLE 5

APPROXIMATE CORRECTION FACTORS FOR CONDENSING STEAM PLANTS[21]

Initial Conditions:

Pressure, psi	400	600	850	1250	2000
Temperature, °F	800	825	900	950	1050
Reheat temp., °F	—	—	—	—	1000

Corrections for:

Pressure, Btu/psi	3.3	2.3	0.7	0.55	0.3
Initial temp, Btu/°F	4.3	3.8	3.0	3.0	1.3
Reheat temp, Btu/°F	—	—	—	—	1.2
Vacuum, Btu/in. Hg	320	220	200	200	200
Feed temp, Btu/°F	1.5*	1.4*	1.1*	0.8*	1.5†

*Change in boiler efficiency is not included.
†Change in boiler efficiency is included.
Note: Corrections represent the change in net plant heat rate (Btu per net kwhr) per unit change in variable.

Alternative Methods of Central Station Energy Production

It has been noted that increased thermal efficiencies from fossil fueled steam plants have reached a plateau from which they are not likely to rise dramatically. In addition, it has been noted that high temperature gas and molten salt breeder nuclear reactors will increase their efficiency only slightly. Therefore, it might be

TABLE 4

PROJECTED FRESHWATER REQUIREMENTS FOR STEAM ELECTRIC POWER GENERATION—1965 - 2020[20]

	Million Gallons/Day			
	1965	1980	2000	2020
Condenser Requirement	84,800	216,400	477,700	806,400
Withdrawn*	68,500	141,500	267,200	419,600
Consumptive Use	758	1,760	4,645	8,110

*Deficit between requirement and withdrawn made up by recirculation.

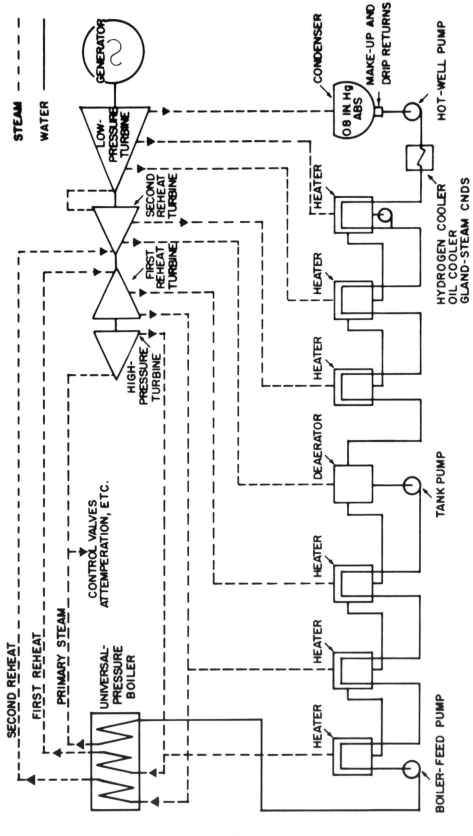

FIGURE 2

Simplified Schematic Heat Balance Diagram for a Super Critical Double Reheat Cycle

16

FIGURE 3
Temperature-Entropy Diagram

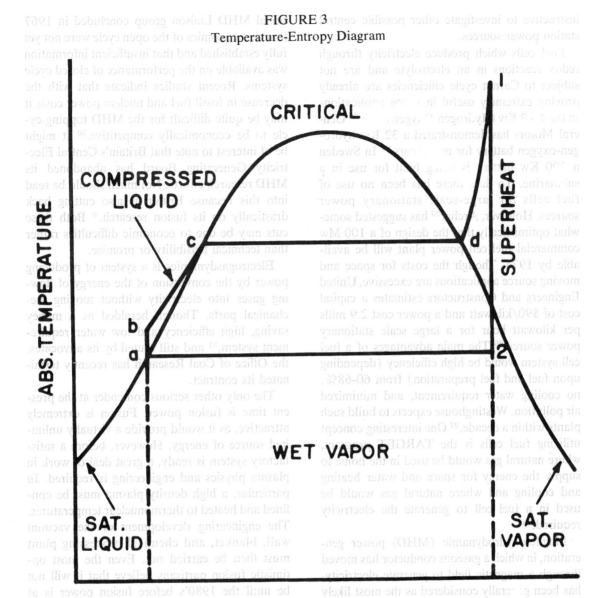

FIGURE 3
Temperature-Entropy Diagram

instructive to investigate other possible central station power sources.

Fuel cells which produce electricity through redox reactions in an electrolyte and are not subject to Carnot cycle efficiencies are already proving extremely useful in space applications in the 1 - 2 Kw (Hydrogen-Oxygen) size.[23] General Motors has demonstrated a 32 Kw hydrogen-oxygen battery for motor cars.[24] In Sweden a 200 Kw battery is being built for use in a submarine. To date there has been no use of fuel cells for large-scale stationary power sources. However, Archer[25] has suggested somewhat optimistically that the design of a 100 Mw commercial fuel cell power plant will be available by 1974. Though the costs for space and moving source applications are excessive, United Engineers and Constructors estimates a capital cost of $90/kilowatt and a power cost 2.9 mills per kilowatt hour for a large scale stationary power source.[26] The main advantages of a fuel cell system would be high efficiency (depending upon fuel and fuel preparation) from 60-68%, no cooling water requirement, and minimized air pollution. Westinghouse expects to build such plants within a decade.[27] One interesting concept utilizing fuel cells is the TARGET program, where natural gas would be used in the home to supply the energy for space and water heating and cooling and where natural gas would be used in a fuel cell to generate the electricity required.[28]

Magnetohydrodynamic (MHD) power generation, in which a gaseous conductor has moved through a magnetic field to generate electricity, has been generally considered as the most likely new system for use in the central power station. Though many engineering problems remain to be solved, the Soviet Union is building a 75 Mw pilot plant in Moscow and a 20 Mw unit is installed at the Tullahoma Arnold Engineering Development Center for hypersonic wind tunnel use. MHD systems with single cycles can achieve thermal efficiencies of 50-55%, which could possibly be boosted to 60-70% by using a binary cycle.[29] If a binary cycle using a gas turbine is utilized, the thermal pollution problem is avoided entirely. However, even the most ardent advocates of MHD agree central station use is at least 10 years away, and the Interna-

tional MHD Liaison group concluded in 1967 that the economics of the open cycle were not yet fully established and that insufficient information was available on the performance of closed cycle systems. Recent studies indicate that with the decrease in fossil fuel and nuclear power costs it may be quite difficult for the MHD topping cycle to be economically competitive.[30] It might be of interest to note that Britain's Central Electricity Generating Board has abandoned its MHD research efforts. Not much should be read into this because Britain is also cutting back drastically on its fusion research.* Both these cuts may be due to economic difficulties rather than technical feasibility or promise.

Electrogasdynamics is a system of producing power by the conversion of the energy of flowing gases into electricity without moving mechanical parts. Though heralded as a money saving, high efficiency and low water requirement system,[31] and still touted by its advocates, the Office of Coal Research has recently terminated its contract.

The only other serious contender at the present time is fusion power. Fusion is extremely attractive, as it would provide a virtually unlimited source of energy. However, before a satisfactory system is ready, a great deal of work in plasma physics and engineering is required. In particular, a high density plasma must be confined and heated to thermonuclear temperatures. The engineering development of the vacuum wall, blanket, and chemical processing plant must then be carried out. Even the most optimistic fusion partisans believe that it will not be until the 1980's before fusion power is at the state of fission power today. In addition, fusion power will still be limited by Rankine cycle efficiencies though they may be higher than present day steam plants through the use of topping cycles.

Direct use of power from the sun has also been investigated,[32] but the near term use in central station power plants is not promising. The role of thermal electricity and thermionics in central station usage seems to be limited.

Therefore, we may conclude that the amount of waste heat to be released will continue to grow and the hopes of any great reduction by increased efficiency will be negated by the over-

*Some British scientists are working cooperatively with Soviet scientists on fusion research (Editor's note.)

all rise in the utilization of electrical power.

The possibilities of dilution must be considered next. The average flow of water in U. S. rivers and streams is about 440 trillion gallons per year. It does not seem likely that there is any great possibility of substantially increasing this amount. The NAS-NRC Report, *Weather and Climate Modification—Problems and Prospects,* has indicated that artificial seeding has produced some evidence for precipitation increases of as much as 10 or even 20% over areas as large as 1000 square miles over periods ranging from weeks to years.[33]

A twenty percent increase in the flow of all rivers while welcome and useful would not, however, solve the problem. It would also be possible through the use of low flow augmentation techniques, etc., to even out the flow during the course of the year, but this still would not solve the problem. Therefore, since technological help in the form of increased efficiencies in energy utilization or by major additions to dilution water are remote, one must turn to a means of dissipating the heat directly to the atmosphere and to restricted water bodies, rather than to streams and reservoirs.

Alternative Methods of Heat Dissipation

It has already been shown that there exists little likelihood of such increases in thermal efficiencies in conventional power sources or of practical application of new power sources that the major portion of heat generated will be effectively used. It has also been shown that the possibilities of increased diluent water flow through weather modifications or low flow augmentation are such that disposal to the atmosphere will be the major methodology to be utilized in the future.

In the water short regions of the country, this is already the case. In the Colorado region the use of cooling towers and ponds is already well established. Nineteen out of twenty-two plants having a capacity of 100 megawatts or greater are already using these auxiliary cooling devices. Already, in 1965, over 22.5% of the plants greater than 100 megawatts in the nation used cooling ponds or towers.[34]

PHYSICAL, BIOLOGICAL, AND CHEMICAL EFFECTS
ON WATER QUALITY

Physical

Temperature affects nearly every physical property of concern in water quality management including density, viscosity, vapor pressure, surface tension, gas solubility, and gas diffusion. Table 6 (see page 20) details the change in these properties with change in temperature.

The solubility of oxygen is probably the most important of these parameters, inasmuch as dissolved oxygen in water is necessary to sustain many forms of aquatic life. Obviously, the lower solubility induced by higher temperatures, if combined with an organic load and an increased bacterial respiration rate, could lead to such low levels of oxygen that fish could not survive.

Since vapor pressure is a driving force in evaporation, an increase in temperature will cause an increase in evaporation, assuming other factors to be constant. Evaporation is one of the key mechanisms in cooling water bodies. It also follows that increased temperature may lead to increased sedimentation in the receiving water, thus, possibly adding to the sludge deposit problem. Because of the changes in the properties of water with temperature, changes in filtration, flocculation, and ion exchange will also occur.

The increased temperature alone may also be a problem. McKee and Wolf[35] report that water with a temperature of 10°C is usually satisfactory and waters with temperatures of 15°C or higher are usually objectionable for drinking purposes.

The effect of temperature on the reaeration coefficient, k_2, can be very important in determining the waste assimilative capacity of streams below, above, and in an impoundment. The reaeration coefficient increases monotonically with temperature as shown by Streeter.[36]

$$k_{2(T°C)} = k_{2(20°C)} 1.016^{(T°C-20°C)} \quad (1)$$

The precise mechanism by which temperature affects the reaeration rate constant is somewhat uncertain. O'Connor and Dobbins[37] consider the effect to be through the change in the molecular diffusion coefficient of oxygen in water. Krenkel and Orlob[38] proposed that the effect was caused by the greater proportion of oxygen molecules having the energy required to create a "site" in the water surface at higher temperatures. They noted that the requisite activation energy was about 5000 calories per mole, which is close to that for the flow of water. Because the water and oxygen molecules are approximately the same size, this result would be expected.

Biological

This section discusses temperature changes as they relate to purification processes only in the receiving water. According to Gunter,[39] temperature is the most important single factor governing the occurrence and behavior of life. Temperature effects on microorganisms are significant to the biological processes of waste stabilization because of induced changes in growth rates and changes in death rates.

In general, the higher the temperature, the more active a microorganism becomes, unless the temperature or a secondary effect becomes a limiting factor. Thus, metabolic activity of thermophilic organisms is much greater at their optimum than psychrophilic organisms at their optimum. Examination of the known effects of temperature on waste treatment processes demonstrates the validity of this statement. It should be noted that a distinct difference exists between the ability of a microorganism to endure a given temperature and its ability to grow well under identical conditions.

The organism of particular interest to water quality management is *Escherichia coli*, as it is the prime indicator of fecal pollution. It should be noted that increased temperatures may lead to optimum growth conditions for this organism in receiving waters.

The rate constant, k_1, which is conventionally used to describe the rate at which biological oxidation takes place in laboratory investigations is defined as:

$$\frac{dL}{dt} = -k_1 L \quad (2)$$

where L = Biochemical Oxygen Demand (BOD),
t = time, days,
k_1 = rate constant, days^{-1}.

Temperature has a profound effect on the rate of oxidation. In the usual receiving water, a multitude of different organisms is active in waste assimilation, all with their own characteristics and temperature tolerances. The distribution of organisms may change drastically

TABLE 6

WATER PROPERTIES AS A FUNCTION OF TEMPERATURE

Temperature °C	Vapor Pressure mmHg	Viscosity Centipoise	Density g/ml	Surface Tension Dynes/cm	Oxygen Solubility mg/l	Oxygen Diffusivity ft²/hr	Nitrogen Solubility mg/l
0	4.579	1.787	0.99984	75.6	14.6		23.1
5	6.543	1.519	0.99997	74.9	12.8		20.4
10	9.209	1.307	0.99970	74.2	11.3	61	18.1
15	12.788	1.139	0.99910	73.5	10.2	71	16.3
20	17.535	1.002	0.99820	72.8	9.2	81	14.9
25	23.756	0.890	0.99704	72.0	8.4	92	13.7
30	31.824	0.798	0.99565	71.2	7.6	106	12.7
35	42.175	0.719	0.99406		7.1		11.6
40	55.324	0.653	0.99224	69.6	6.6		10.8

20

with shifts in temperature or types of waste, each species having a different rate of metabolism. Because the composite metabolism includes many coupled reactions, each with its own characteristics, the composite rate-limiting steps may shift with temperature changes, thus precluding the interpretation of the process as a single reaction. In addition, the increasingly rapid rates of protein denaturation, which takes place at temperatures above 30°C, must be considered. The protein portion of the enzyme is inactivated and the reactions apparently begin to slow down at temperatures above 30°C. It is, therefore, obvious that, at present, a semi-empirical relationship must be used to describe the variation of the BOD reaction rate with temperature. Such a relationship proposed by Streeter[40] and still favored by most workers is

$$k_{1(T)} = k_{1(20)} \, \theta^{(T-20)} \qquad (3)$$

in which $k_{1(T)}$ = the BOD rate constant at temperature T, in °C; $k_{1(20)}$ = the rate constant at 20°C; and θ = a constant, usually taken to be 1.047. This formulation was originally expected to hold for temperatures between 2°C and 40°C. However, Gotaas[41] showed that the maximum reaction rate is reached at about 30°C, and decreases at higher temperatures. He also found that the value of the constant, θ, is higher at lower temperatures. Both of these findings can be rationalized. The decrease in reaction rate at high temperatures could be caused by thermal inactivation of enzymes, and the high rate changes at lower temperatures could be caused by the shift from a psychrophylic to a mesophylic population predominance. Gotaas found that the variation of reaction rate with temperature could be accurately described by Eq.3., with three temperature ranges. They were

$$5 \text{ to } 15°C \quad \theta = 1.109$$
$$15 \text{ to } 30°C \quad \theta = 1.042$$
$$30 \text{ to } 40°C \quad \theta = 0.967$$

For the range 5 to 30°C, the average value of θ was 1.071.

It should not be expected that the data of different workers on this subject will agree. It should be obvious, in the light of the discussion of the BOD reaction herein, that the rate of oxidation will vary with the type of waste and the type of biological population which grows in response to and oxidizes the waste. Under differing conditions, different biological populations will vary in their temperature susceptibility so that temperature coefficients and temperatures of maximum reaction rates will also vary.

It should be noted that Schroepfer et al.[42] found different results using stabilized effluents and river waters than previously reported. Also, their work assumed that at elevated temperatures of 35°C, significant nitrification began in less than five days.

Zanoni[43] found that BOD data at all temperatures more nearly followed a second order reaction equation and that the second order equation ultimate demands were higher than those predicated by the first order equation. He also found values for θ between 10 to 30°C similar to those of past investigations. In a subsequent study Zanoni[44] further substantiated past work on both the ultimate demand and the temperature constants for the carbonaceous stage. However, he reported that the nitrogenous stage demonstrated an optimum rate at 22°C and decreased until 30°C was reached. For the nitrogenous phase, he found a value of θ of 1.097 for temperatures from 10°C to 22°C, and 0.877 for temperatures from 22°C to 30°C.

Temperature also has a profound effect on the bactericidal and viruscidal activity of chlorine compounds. Butterfield[45] reported that a 20°C reduction in temperature requires nine times the exposure period or 2.5 times as much chloramine to produce a 100% kill. In an earlier paper, Butterfield[46] reported similar results for chlorine. Similar studies with viruses have indicated that a 10°C decrease in temperature will increase the time required for a given amount of viral destruction by chlorine by 200 to 300%.[47]

Chemical

An increase in temperature will usually have a profound effect on chemical reactions, the rate of reaction being approximately doubled for each 10°C rise in temperature.[48] The variation of an equilibrium constant, k, may be expressed as:

$$\frac{d (\ln k)}{dT} = \frac{E_a}{RT^2} \qquad (4)$$

where:

T = absolute energy °K
E_a = activation energy calories/mol
R = gas constant calories/°K mol

Integration of Equation (4) between temperatures T_1 and T_2 yields:

$$\ln\frac{k_2}{k_1} = \frac{E_a(T_2\text{-}T_1)}{RT_1T_2} \quad (5)$$

The activation energy, E_a, is a measure of the requisite energy for a molecule to take part in a reaction and for many reactions may be determined by plotting ln k against 1/T, the slope of the line being $-E_a/R$.

The following are other methods common to the pollution control field for expressing the temperature dependence of reaction rates:[49]

$$\frac{k_2}{k_1} = \theta^{T_2\text{-}T_1} \quad \text{where } \theta = \frac{k_2}{k_1} \text{ for } T_2\text{-}T_1 = 1 \quad (6)$$

and

$$Q_{10} = \frac{k_2}{k_1} \text{ for } T_2\text{-}T_1 = 10 \quad (7)$$

It should be noted that T_2 and T_1 are measured in degrees Centigrade for Equations 6 and 7 and that Q_{10} and θ vary with temperature. It is also of interest to note that $Q_{10} = \theta^{10}$.

Temperature affects not only the rate at which a reaction occurs, but the extent to which the reaction takes place. When considering temperature changes in a receiving water, one must contemplate changes in ionic strength, conductivity, dissociation, solubility, and corrosion. With an increase in temperature, these changes might very well result in differing chemical requirements in the water treatment plant.

EFFECTS OF HEATED DISCHARGES ON WASTE ASSIMILATION

A major factor affecting waste assimilation capacity is the temperature of the receiving waters and the thermal additions to that water. When a heated discharge is injected into a receiving water, it may result in a stratified flow condition.

Stratification

The profound influence of stratified flow on the waste assimilative capacity of a receiving water is of particular interest to the stream analyst. The major effects to be considered are the following:

1. Mixing between the upper and lower layers is inhibited. Thus, oxygen replacement through the natural forces of stream self-purification at the water surface cannot take place in the lower layer of water.

2. Because of the lack of mixing, organic wastes discharged into the lower layer are not allowed the benefit of that portion of the stream flow in the upper layer. This results in less oxygen, less dilution, and a more concentrated organic load exertion on the natural stream self-purification process.

3. The rate of oxygen depletion in the lower layer may be increased because of the increased concentration of organic material.

4. The waste assimilative capacity will be reduced because of the decrease in the ability of the water to hold dissolved oxygen, and an increase in the metabolic activity of stream organisms, and an increased rate of biochemical oxygen demand exertion or oxygen depletion.

Because of these combined effects of increased temperatures, it is obvious that the addition of heated discharges to a receiving water, just as sewage or other organic waste materials, should be considered as a pollutant. Senator Muskie has stated[50] that: "It is the opinion of the Senate Subcommittee on Air and Water Pollution that excessive heat is as much a pollutant as municipal wastes or industrial discharges."

Waste Assimilative Capacity

An example of the change in river waste assimilation capacity caused by both temperature and impoundment was presented by Krenkel et al.[51] A paper mill of the Georgia Kraft Company, which had previously satisfactorily discharged its waste effluent into the free-flowing Coosa River, must now discharge into the backwater of a downstream impoundment and receive its flow as regulated by a peaking hydroelectric power plant upstream. The daily fluctuation in discharge may range from 200 to 7800 cfs. In addition, a steam electric generating plant, with a capacity of 300,000 kilowatts, has located adjacent to the mill, and the condenser water raises the temperature of the stream several degrees.

The effects of temperature on the stream self-purification process is demonstrated by Figure 4 (see page 24) which shows the variation of the rate constants k_1 (de-oxygenation) and k_2 (reaeration) with respect to temperature. Examination of this relationship demonstrates that an increase in temperature causes a considerable increase in k_1. While k_2 also increases with increasing temperature, it is negated by the combination of a lesser dissolved oxygen content and a greater rate of change of k_1 with temperature.

The overall effects of the impoundment on the rate of oxygen recovery is demonstrated by the lower curve, which depicts the reaeration rate constant under existing, impounded conditions. Note that, while k_1 at a given temperature is unchanged, the value of k_2 at any temperature is significantly reduced.

In order to illustrate the effect of temperature on the waste assimilative capacity of the Coosa River, field data were used to obtain Figure 5 (see page 25) which depicts the oxygen balance in the Coosa River prior to the previously mentioned water resources developments. Note that under these conditions, the Coosa River easily assimilated 28,000 pounds per day of BOD at the existing river temperature of less than 25°C. Even under the free-flowing conditions, however, a temperature of 30°C would cause the dissolved oxygen level to fall below 4.0 mg/1, the minimum required to satisfy the existing stream standards.

Figure 6 (see page 26) is presented to show that, because of the combined effects of the water resources developments and the observed temperature increases, the Coosa River can no longer satisfactorily assimilate the 28,000 pound BOD load, as under the previously existing free-flowing conditions.

The quantitative effect of temperature on the waste assimilative capacity of the Coosa River can be shown on Figure 7 (see page 27) where the possible waste load that would not deplete the dissolved oxygen content of the river below 4.0 mg/1 is plotted versus the river water temperature.

It may be concluded from this figure, that if the river temperature were 25°C, a 5°C increase in temperature under free-flow conditions is equivalent to 11,000 pounds per day of BOD and a 5°C increase in temperature under the existing, impounded conditions is equivalent to 5200 pounds per day of BOD.

Note the significant reduction in assimilative capacity already caused by the impoundment and the observed increase in temperature, as demonstrated by comparing the curve for free-flow conditions and the curve for impounded conditions. It is obvious from these curves, which were computed from observed conditions, that the addition of heat and the impoundment of the Coosa River have had the same end result as if an equivalent amount of sewage or other organic waste material were added to the river.

Because of the previously described reduction in the waste assimilative capacity of the Coosa River, the mill has reduced its waste load to the river to less than 20,000 pounds of BOD per day which is the allowable load permitted by the Georgia Water Quality Control Board in order to maintain an adequate dissolved oxygen content.

Bohnke[52][53] states that an increase in temperature of the river water of 6°C will cause a 6% reduction in self-purification capacity at a minimum oxygen content of 3 mg/1. The loss of assimilative capacity due to a specific power plant on the Lippe was said to be 10,000 population equivalents at low flow.

A recent report by Drummond[54] estimates that the Chattahoochee River in Georgia can assimilate over three times as much organic material at 20°C as it can at 30°C without excessive depletion of oxygen in the water. The city of Atlanta is quite concerned because of two steam electric generation plants, which, when operating at full load, will raise the maximum average daily river temperature from 20°C

FIGURE 4
Self-Purification as a Function of Temperature

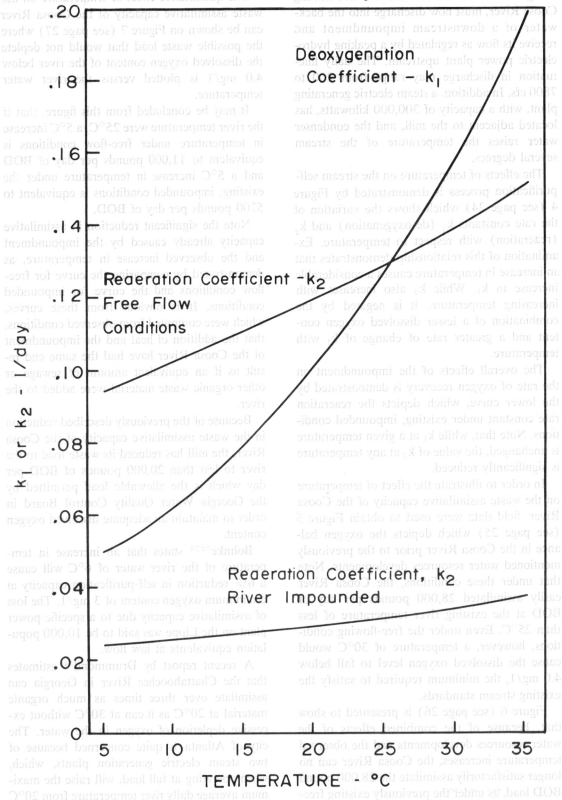

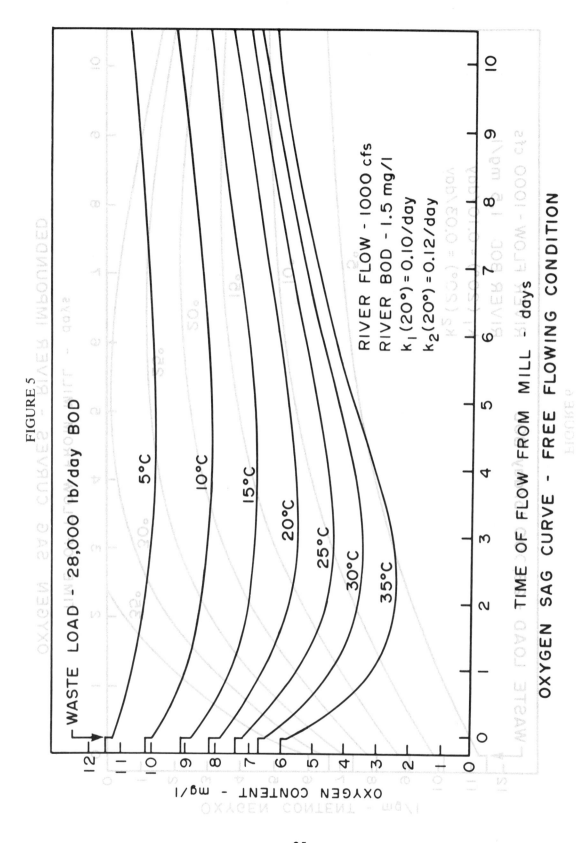

FIGURE 5

WASTE LOAD – 28,000 lb/day BOD

5°C
10°C
15°C
20°C
25°C
30°C
35°C

RIVER FLOW – 1000 cfs
RIVER BOD – 1.5 mg/l
$k_1(20°) = 0.10$/day
$k_2(20°) = 0.12$/day

OXYGEN CONTENT – mg/l

TIME OF FLOW FROM MILL – days

OXYGEN SAG CURVE – FREE FLOWING CONDITION

FIGURE 6

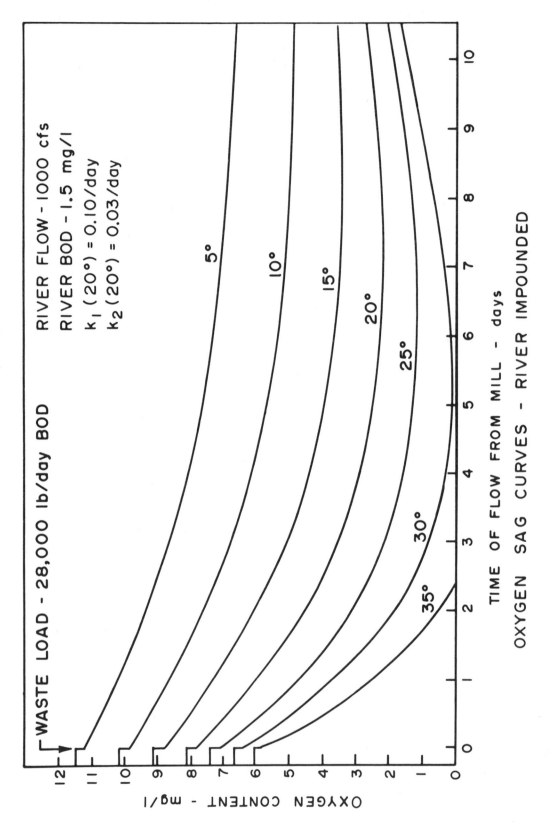

WASTE LOAD - 28,000 lb/day BOD

RIVER FLOW - 1000 cfs
RIVER BOD - 1.5 mg/l
$k_1 (20°) = 0.10$/day
$k_2 (20°) = 0.03$/day

OXYGEN CONTENT - mg/l

TIME OF FLOW FROM MILL - days

OXYGEN SAG CURVES - RIVER IMPOUNDED

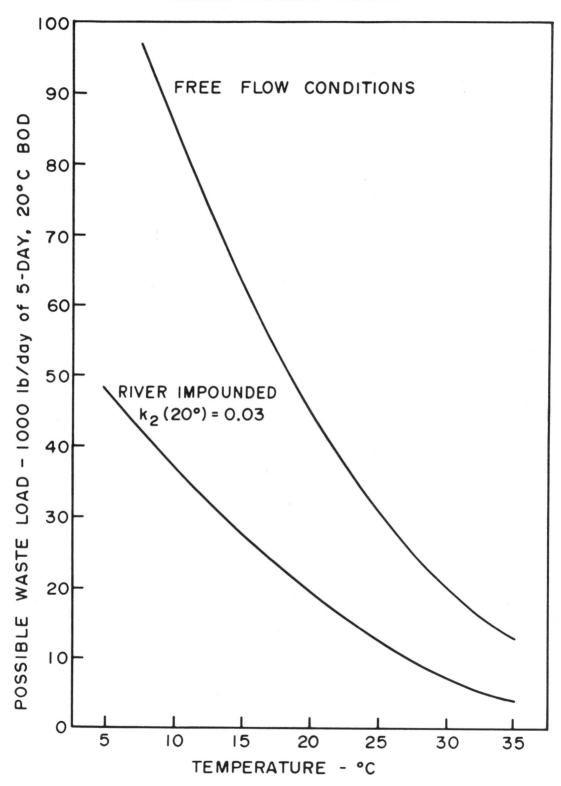

FIGURE 7
Permissible Waste Load at 4 PPM D. O.

to 34°C at minimum flow. It was concluded that, in order to maintain satisfactory dissolved oxygen concentrations in the river during minimum flow, either the steam plant operations must be curtailed, re-regulation of the river must be practiced, or cooling towers must be installed.

It may be concluded from these studies that the addition of heated water to a receiving water can be considered equivalent to the addition of sewage or other organic waste material, since both pollutants may cause a reduction in the oxygen resources of the receiving water.

BENEFICIAL EFFECTS OF HEAT ADDITIONS

Though the predominant results of heating up our natural waters are detrimental to the propagation of game fish and reduce the assimilative capacity of a stream, there are some beneficial effects due to the warmed waters. The total amount of waste heat produced per year, however, is so great that even if all the suggestions for use of waste heat for beneficial use were adopted, the excess heat remaining would still be a problem. For example, the waste heat from power generation would be more than sufficient to heat every home in America. In the following sections the existing and proposed beneficial uses are listed.

District Heating

Possibly the most advanced use of low grade and waste heat is for district heating.[55] This has been used in Finland for the 20,000 inhabitant city of Tapiola Garden since 1953. The English also have used waste heat for district heating.

Increased Biological Production

Though it has been observed that on the Potomac River and many other rivers fishermen tend to congregate in the winter by the heated pools below power stations for better fishing, it has also been noted that fishing there is poorer in the summer.[56]

Research into the possible increased growth of fish below steam electric generating plants has been underway since at least 1953. However, definitive statements are still not possible. In 1953, the Central Electricity Generating Board determined the rate of growth of flounder at the Newton Abbot Generating Station.[57] Some of the earlier work had dealt with the possibility of utilizing increased yields of algae for food material and talked of farms below the sea.[58] In 1962 the experiments were extended to feed the algae to clams to help solve the problems of harvesting the algae and to ease the preparation of the final product.[59] The economics of the production have not been completed.

Mihursky[60] has suggested a variety of possible constructive uses of thermal additions to estuaries, including a complete recycling of organic wastes from humans to sewage treatment plants to inorganic fertilizers to algae, zooplankton, shellfish, and fish to food processing to people to organic wastes, etc. Utilization of the waste heat could be made at the sewage treatment plant, in the enhanced growth of algae, zooplankton, shellfish and fish, and district heating. Studies of individual sections of this cycle are being carried out.

Shrimp farming in the effluent of the Turkey Point Power Plant is being studied as is pompano cultivation.[61] Lobsters may also benefit from heated waters. It has been suggested that the lower yields of lobster in recent years may be due to cooler waters in their breeding grounds off the Maine Coast. It has been further suggested that the heated discharge from cooling stations could be used to warm shoreline coves to increase lobster yields.[62]

Long Island Lighting Company at Northport, New York, is experimenting with the aquaculture of oysters in the heated discharge from their nuclear power plant.[63] A report from the Electric Boat Division has suggested that it might be possible to raise 18 million kilograms of warm water fishes annually by using thermal discharges. Present annual per capita consumption of fish is approximately 10.5 pounds.[64]

Perhaps the most original suggestion is that heated waters in England be used to raise ornamental fish, some of which sell for over $140 per pound.[65] In addition, Chinese Carp, Tilapia, Grass Carp, Silver Carp, and Rainbow and

Brown Trout have been used in fish culture experiments by England's Central Electricity Generating Board.[66]

Some inadvertent beneficial use of the warmed waters may have already occurred. The growth and spread of the American Hard-shell Clam in English waters may well have been due to the warming effect of cooling water effluents.[67]

The ability of thermophilic algae to survive at temperatures as high as 85°C has been reported. Recent studies, however, indicate that cyanidium will not undergo cell division at temperatures much greater than 60°C.[68] It is possible, therefore, for some species to survive very high temperatures. At Yellowstone Park the optimal temperature for algae development in the alkaline hot springs was 51° to 56°C.[69]

Control of Marine Fouling

The reversal of the flow of cooling sea water streams through intake pipes in order to raise the temperature to limit the creatures inhabiting the pipes is not new. The use of heated waters for the control of mussels for a station on the coast of California is shown to be less expensive than the use of chlorine.[70] A follow-up of the original work 10 years later shows that thermal control was still the most successful as well as the most practical means of controlling fouling.[71]

Water Works Treatment

Many studies have been made on the effect of temperature on water treatment processes. Renn sums up the experience to 1956[72] by noting that Fair and Geyer[73] find that "the efficiency and effectiveness of flocculation and of filtration of floc-bearing water rises with rising temperatures." Most other authors agree, though Velz[74] disagreed. Camp's studies[75] showed that the optimum condition for flocculation is determined by 3 variables, iron-alum dose, pH, and temperature. Velz's findings may be in agreement with this since he also showed that the isoelectric point of the flocculating system shifts drastically with temperature. Because of this increased efficiency in flocculation, the State of Pennsylvania's Committee of the Effects of Heated Discharges found in 1962 that savings in chemicals for water treatment would be 30 to 50 cents per million gallons for each 10°F rise in temperature.[76] This may be compared with

an average cost of chemicals of $14 for treating water and a range of $3 to $30 per million gallons.

Waste Treatment

There is a considerable body of literature on the effects of heating sludge, but little information on the effects of temperature greater than 60°C. Recently, raw sewage has been heated to 100°C to determine if there were any improvements in settling efficiency.

Crotty et al.[77] found that heat treated wastes were more homogeneous, and settled faster. Further work is required. Earlier German work cited by Crotty on preheating sewage and raw sludge to 100° to 120°C shows that they will be more amenable to sedimentation and biological treatment.

Irrigation

The intake of the Oroville Power Plant on the Feather River was especially constructed at considerable extra expense to draw water from various levels to avoid damage to downstream agricultural interests by excessively cool water.[78] The same results could have been achieved by mixing the heated waters from a thermal power station with the discharge from the hypolimnion of a storage reservoir.

Ice-free Shipping Lanes

A recent study by Dingman et al.[79] showed that it should be possible to keep significant portions of the Saint Lawrence Seaway open the year around by the judicious location of central station electric power complexes. This would save transportation costs of several million dollars per year. It is estimated that a 600 Mw_e reactor could keep a stretch of the river between 11 and 16 miles ice-free. No study was made of the ecological effect of such an undertaking, however.

Water and Sediment Discharge

Since the viscosity and density of water decrease with temperature, a change in temperature should have an effect on both sediment transport and water flow. It had previously been thought that the variation of only a few degrees would have a small effect.[80] More recent data, however, suggests that these effects may be quite

29

important.[81] Colder water is more viscous than warmer water and therefore has a higher carrying capacity for sediment than does warmer water. In warmer water, coarser material settles out and so for any crossing in the river (which acts as a submerged weir) there is less flow past that cross-section for the same stage. Decreasing water temperature may be assumed to make a difference of 10 to 20% in expected discharge for a given stage.[82] Sudden rises in temperature may cause such a deposition of coarse material to ground ships following sailing lines.

Studies by Colby and Scott[83] indicated that changes in temperature affect bed material discharge in complex ways. The thickness of the laminar sublayer is changed with the temperature, but it is usually only a small effect. The vertical distribution of suspended material is greatly affected by temperature changes and indicates an approximate doubling of bed material discharge when temperatures drop from 80 to 40°F, assuming that the mean velocities, depths, and sizes of bed sediments remain constant. Finally, the bed configuration and, therefore, the resistance to flow are affected by changes in temperature. Their effect may be large or small. Colby,[84] in an earlier summary of fluvial sediments, had noted that the fall velocities of sediment particles increase with an increase in water temperature. For particles larger than 1 mm, however, the percentage increase was small. For sediments in the size range 0.1 to 0.4 mm, fall velocity, vertical distribution, and discharge of sediments were greatly affected by temperature changes.[85] Franco,[86] in a more recent study, confirms some of these results and found that the effects of water temperature on bed load appeared to be mostly in the formation of the bed and bed roughness.

Water Supply Source

A serendipitous benefit was obtained in Henderson, Kentucky, when it was found necessary to replace the town's water supply system. Over one quarter of a million dollars was saved by the use of the condenser cooling waters as the intake to the town's water supply system.[87] Fortunately, no chemicals were added to the intake water and the warmer water had not to that date caused any problems.

PREDICTION OF HEAT DISSIPATION

Obviously, the simplest method of disposing of heated waste waters is to discharge them directly to the receiving water and then allow natural forces to bring the water back to an equilibrium temperature. This process is known as once-through cooling and is shown schematically on Figure 8 (see page 31). In order to predict the behavior of these heated effluents, it is necessary to resort to an energy balance.

Much of the work on the heat balance has been a result of the interest in evaporation losses by hydrologists and hydraulic engineers and is therefore primarily concerned with the prediction of evaporation rates. A brief presentation of these studies will be made. For a more extensive literature review of this subject, the reader is referred to Anderson,[88] U. S. Geological Survey,[89] U. S. Public Health Service,[90] and Edinger and Geyer.[91]

The Energy Balance

The use of the energy balance approach for estimating evaporation, elucidated by Angstrom,[92] was used by Schmidt[93] to approximate ocean evaporation in 1915 and has since been applied to compute evaporation from water bodies of all sizes. The relatively recent development of more sophisticated instrumentation has allowed the energy budget to be utilized with a fair degree of reliance.

The energy budget was first tested against a water budget control at Lake Hefner in 1950-1951.[94] It was concluded that satisfactory results were obtained for periods of ten days or more. A second check against a water budget was made at Lake Colorado City, Texas, in 1954-1955, where inflow was extremely small and outflow was zero.[95] The components of the energy budget per unit surface area of a reservoir per unit time may be written as follows:[96]

FIGURE 8
Steam Electricity Generating Station Cooling System

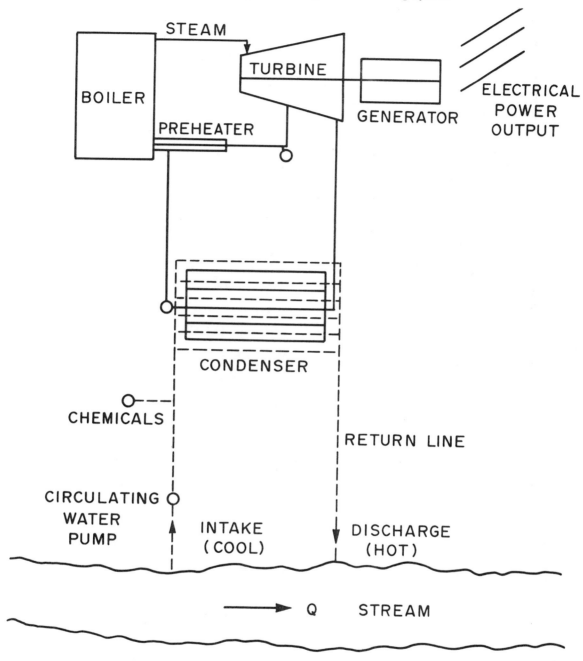

FIGURE 9

MECHANISMS OF HEAT TRANSFER
ACROSS A WATER SURFACE

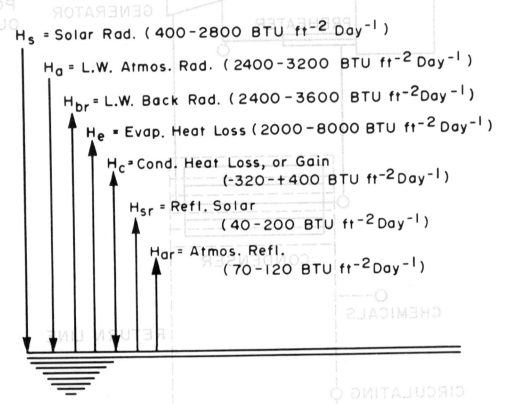

H_s = Solar Rad. (400-2800 BTU ft^{-2} Day^{-1})

H_a = L.W. Atmos. Rad. (2400-3200 BTU ft^{-2} Day^{-1})

H_{br} = L.W. Back Rad. (2400-3600 BTU ft^{-2} Day^{-1})

H_e = Evap. Heat Loss (2000-8000 BTU ft^{-2} Day^{-1})

H_c = Cond. Heat Loss, or Gain
      (-320 - +400 BTU ft^{-2} Day^{-1})

H_{sr} = Refl. Solar
      (40-200 BTU ft^{-2} Day^{-1})

H_{ar} = Atmos. Refl.
      (70-120 BTU ft^{-2} Day^{-1})

NET RATE AT WHICH HEAT CROSSES WATER SURFACE

$$\Delta H = (H_s + H_a - H_{sr} - H_{ar}) - (H_{br} \pm H_c + H_e) \; \text{BTU ft}^{-2} \text{Day}^{-1}$$

H_R	Temp. Dependent Terms
Absorbed Radiation Independent of Temp.	$H_{br} \sim (T_s + 460)^4$
	$H_c \sim (T_s - T_a)$
	$H_e \sim W (e_s - e_a)$

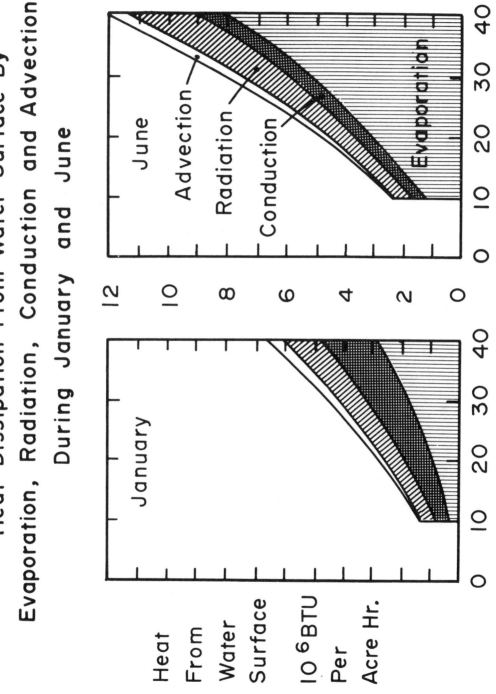

FIGURE 10

Heat Dissipation From Water Surface By
Evaporation, Radiation, Conduction and Advection
During January and June

$$Q_s - Q_r + Q_a - Q_{ar} - Q_{bs} + Q_v - Q_e - Q_h - Q_w = Q \tag{8}$$

where:

Q_s = short-wave radiation incident to the water surface;

Q_r = reflected short-wave radiation;

Q_a = incoming long-wave radiation from the atmosphere;

Q_{ar} = reflected long-wave radiation;

Q_{bs} = long-wave back radiation emitted by the body of water;

Q_v = net energy brought into the body of water in inflow; including precipitation, and acccounting for outflow;

Q_e = energy utilized by evaporation;

Q_h = energy conducted from the body of water as sensible heat;

Q_w = energy carried away by the evaporated water;

Q = increase in energy stored in the body of water.

Edinger and Geyer[97] have depicted the heat transfer terms across a water surface as shown on Figure 9 (see page 32) noting temperature dependent terms and typical values. The addition of heated water discharges simply superimposes the heat addition upon the natural dissipations and additions of energy.

Figure 10 (see page 33) shows the results of calculations by Bergstrom for a water surface in central Illinois.[98] The data demonstrate the relationship of rate of heat dissipation to the elevation of the water surface temperature over natural temperature and the mechanisms by which this dissipation is achieved. It is significant to note that the rate of heat dissipation for a given rise in temperature is greater in summer than in winter and that the heat dissipation by evaporation is much greater in summer than in winter. These calculations would appear to support the contention that heated effluents should be discharged in the most concentrated form possible (neglecting any biological or other effects) in order to dissipate the heat most rapidly.

While the terms in the energy budget are discussed in detail by Anderson[99] and Edinger and Geyer,[100] brief comments pertaining to their origin and variability are in order.

Short Wave Radiation, Q_s

Short wave radiation originates directly in the sun, although the energy is depleted by absorption by ozone, scattered by dry air, absorbed and scattered by particulates, and absorbed and scattered by water vapor. It varies with latitude, time of day, season, and cloud cover. Thus, while this quantity can be empirically calculated, it is much better to measure it with a Pyroheliometer which will give the accuracy required for the energy budget.

Long Wave Atmospheric Radiation, Q_a

Long wave atmospheric radiation depends primarily on air temperature and humidity and increases as the air moisture content increases. It may be a major input on warm cloudy days when direct solar radiation approaches zero. It is actually a function of many variables, including carbon dioxide and ozone, although it can be fairly accurately calculated by means of an empirical formulation as demonstrated in the Hefner studies. Although it can be measured with the Gier-Dunkle Flat Plate Radiometer, it is more convenient to calculate than measure Q_a.

Reflected Short Wave and Long Wave Radiation, Q_r and Q_{ar}

Solar reflectivity, R_{sr}, is more variable than atmospheric reflectivity, R_{ar}, inasmuch as the solar reflectivity is a function of sun altitude and cloud cover, while atmospheric reflectivity is relatively constant. The Lake Hefner studies demonstrated the atmospheric reflectivity to be approximately 0.03, while on an annual basis, the solar reflectivity was 0.06. The Hefner studies used the equation:

$$R_{sr} = a S_a^b \qquad (9)$$

to determine solar reflectivity, where S_a is the sun altitude in degrees and a and b are constants depending on cloud cover. Note that $R_{sr} = Q_r/Q_s$ and $R_{ra} = Q_{ar}/Q_a$.

Long Wave or Back Radiation, Q_{bs}

Water sends energy back to the atmosphere in the form of long wave radiation and radiates almost as a perfect black body. Thus, the Stefan-Boltzman fourth power radiation law can be utilized, or:

$$Q_{bs} = 0.97\sigma (T_o + 273)^4 \qquad (10)$$

where

0.97 = emissivity of water
Q_{bs} = long wave radiation in calories/ cm²/day
σ = Stefan-Boltzman constant = 1.171 x 10⁻⁷ calories/cm²/deg⁴/day
T_o = water surface temperature in °C.

All that is required to compute Q_{bs} is the water surface temperature, and a table giving the value of Q_{bs} for any temperature, T_o, which is readily available or computable.

Energy Utilized by Evaporation, Q_e

Each pound of water evaporated carries its latent heat of vaporization of 1054 BTU's at 68°F; thus Q_e is a significant term in the energy budget. The Lake Hefner study was explicitly promulgated for determining correct evaporation relationships and resulted in the following equation:

$$Q_e = 11.4W_x (e_s - e_a) = BTU/ft^2/day \quad (11)$$

which is of the general type of evaporation formula:

$$E = (a + bW_x) \ (e_s - e_a) \qquad (12)$$

where

a, b = empirical coefficients
W_x = wind speed at some elevation, x.mph.
e_a = air vapor pressure mm Hg

e_s = saturation vapor pressure of water determined from water surface temperature, mm Hg
E = evaporation = Q_e/pL, ft/day
p = density of evaporated water, lb/ft³
L = latent heat of vaporization, BTU/lb

Many expressions have been developed for estimating the evaporation rate, the coefficients differing because of variation in the reference height for measurement of wind speed and vapor pressure, the time period over which measurements are averaged, and local topography and conditions. As stated by Edinger and Geyer,[101] "It would also be expected that the coefficients would be much different for rivers and streams than for lakes and might well be dependent on water velocity and turbulence, particularly in the case of smaller rivers."

Energy Conducted as Sensible Heat, Q_h

Heat enters or leaves water by conduction if the air temperature is greater or less than water temperature. The rate of this conductive heat transfer is equal to the product of a heat transfer coefficient and the temperature differential.

A single direct measurement of this quantity is not available and recourse to an indirect method is necessary. The method involves using average figures of air temperature, water surface temperature, and humidity for the period in question and computing the ratio of Q_h to Q_e, which is known as the Bowen Ratio and expressed as:

$$R_B = \frac{Q_h}{Q_e} = \frac{0.61 \, P \, (T_o - T_a)}{1000 \, (e_o - e_a)} \qquad (13)$$

where

P = atmospheric pressure, millibars
T_a = temperature of air, °C
T_o = temperature of water surface, °C
e_o = saturation vapor pressure corresponding to temperature of water surface, millibars
e_a = vapor pressure of air at height at which T_a is measured, millibars

35

Energy Carried Away by Evaporated Water, Q_w

Water being evaporated from the surface is at a higher temperature than the lake water, and thus energy is being removed. Though some believe this term is included in the conductive energy term, it is not. However, it is relatively small and can be readily computed from

$$Q_w = p_e \, c \, E \, (T_e - T_b) = \frac{cal}{cm^2\text{-}day} \quad (14)$$

where

ρ_e = density of evaporated water, g/cm^3
c = specific heat of water, cal/g
E = volume of evaporated water, $g/cm^2/day$
T_e = temperature of evaporated water, $^\circ$C
T_b = base or reference temperature, $^\circ$C

Advected Energy, Q_v

The net energy change per unit area per unit time in the lake water due to water entering and leaving the lake may be computed from the following expression:

$$Q_v = c_{si} \, V_{si} \, \rho_{si} \, (T_{si} - T_b)$$
$$+ \, c_{gi} \, V_{gi} \, \rho_{gi} \, (T_{gi} - T_b)$$
$$- \, c_{so} \, V_{so} \, \rho_{so} \, (T_{so} - T_b)$$
$$- \, c_{go} \, V_{go} \, \rho_{go} \, (T_{go} - T_b)$$
$$+ \, c_p \, V_p \, \rho_p \, (T_p - T_b) \div A \quad (15)$$

in which

Q_v = advected energy, cal cm^{-2} day^{-1}
c = specific heat of water, ($\simeq$ 1 cal g^{-1} deg^{-1})
V = volume of inflowing or outflowing water, cm^3 day^{-1}
ρ = density of water, ($\simeq$ 1 g cm^{-3})
T = temperature of water, $^\circ$C
A = average surface area of reservoir, cm^2.

The subscripts are as follows

si = surface inflow
gi = groundwater inflow
so = surface outflow
go = groundwater outflow
ρ = precipitation
b = base or reference temperature, usually taken as 0°C.

Since some of the terms in Equation 15 may

not be measurable, a water budget is performed for the same period, evaporation is estimated, and the unknown terms are found by trial and error. In many cases the ground water and precipitation terms may be negligible and, therefore, ignored.

Increase in Energy Stored, Q

The change in storage in the energy-budget equation may be either positive or negative, and is found from properly averaged field measurements of temperature and the following equation

$$Q = c\rho_1 \, V_1 \, (T_1 - T_o)$$
$$- \, c\rho_2 \, V_2 \, (T_2 - T_o) \div At \quad (16)$$

in which:

Q = increase in energy stored in the body of water, cal cm^{-2} day $^{-1}$
c = specific heat of water, ($\simeq$ 1 cal g^{-1})
ρ_1 = density of water at T_1 ($\simeq$ 1 g cm^{-3})
V_1 = volume of water in the lake at the beginning of the period, cm^3
T_1 = average temperature of the body of water at the beginning of the period, $^\circ$C
ρ_2 = density of water at T_2, ($\simeq$ 1 g cm^{-3})
V_2 = volume of water in the lake at the end of the period, cm^3
T_2 = average temperature of the body of water at the end of the period, $^\circ$C
T_o = base temperature, $^\circ$C
A = average surface area, cm^2 during the period
t = length of period, days

Discussion

From this necessarily brief discussion of the various parameters comprising the energy-balance, it may be concluded that it is possible to predict heat dissipation using these concepts. Obviously, the reliability of the results will depend on the degree of sophistication used in the theoretical approach and the frequency and accuracy of the measurements taken.

Edinger and Geyer[102] have presented these relationships in some simple curves, enabling one to calculate the values quickly.

However, in most cases one wishes to know the heat dissipation for a plant not yet built and one cannot measure the surface temperatures. Therefore, one must make some assumption on

36

the thermal structure of the body of water and use iteration techniques to determine the water evaporated; in addition, continuity must be maintained.

Heat Dissipation from Reservoirs and Lakes

One of the first attempts to predict the dissipation of heat from cooling ponds was by Lima.[103] While the theoretical aspects of heat transfer are not presented, several graphs allowing estimates of areas required for cooling were empirically derived. Similar, although more sophisticated formulations were presented by Throne.[104] The graphs contained in these papers are quite useful in obtaining rough estimates of cooling pond performances.

The United States Geological Survey has been a pioneer in heat balance studies with several classic publications resulting from rather large field-scale investigations. Probably the most widely quoted study in the literature is the Lake Hefner investigation,[105] which was a comprehensive attempt to compare various methods for evaporation determination. Four methods for the determination of evaporation were examined: the water budget, the energy budget, the use of mass transfer theory, and the use of evaporation pans. In the water budget method, evaporation is determined by measuring inflow, outflow, and changes in storage; however, it is not generally useful because evaporation is often small when compared to errors in measuring inflow, outflow, seepage, and bank storage. The energy budget method involves precise measurement of energy changes within the body of water and the mass-transfer concept relates factors affecting mass transfer of water vapor to the evaporation process. The use of pans is subject to error because the experimentally determined "pan coefficients" apply only for the conditions under which they were determined. An excellent review of these procedures and a discussion of required instrumentation is contained in a pre-study publication by the U.S. Navy.[106]

Because of unusually favorable conditions, the water budget method yielded results good enough to use as a base for comparing the other more practical procedures for evaporation determination, inasmuch as errors in the water budget did not exceed 5% of the monthly evaporation.

The energy budget method showed that for periods over seven days, evaporation could be determined within ± 5% if all terms in the energy budget are carefully evaluated, particularly changes in energy storage.

All of the data collected during the Lake Hefner studies are contained in another report which is useful to determine the types of observations and the order of magnitude of the results for similar studies.[107]

The studies of Lake Hefner were extended to Lake Mead as described in a later report by Harbeck.[108] Both the energy budget and mass transfer methods were utilized with fairly good results. It was found that evaporation could be estimated closely by knowing only water surface temperature, wind speed, air temperature, and humidity.

In order to test the validity of the mass transfer and water budget methods in a humid area, Turner[109] conducted an evaporation study on Lake Michie in North Carolina. He states that the amount of inflow and outflow of the lake, its size, the availability of an accurate stage-area relation, and the amount and variability of net groundwater seepage are the major factors that determine whether or not a lake can be calibrated by using the mass-transfer water-budget method. The method, using the equation used by Harbeck (1962), combines the mass transfer equation

$$E = Nu \, (e_o - e_a) \qquad (17)$$

with the water budget equation

$$\nabla H_A = E + \delta \qquad (18)$$

to obtain

$$\nabla H_A = Nu \, (e_o - e_a) + \delta \qquad (19)$$

where

E = Evaporation, feet
N = Mass transfer coefficient
u = Wind speed, miles per hour
e_o = Saturation vapor pressure, millibars, corresponding to temperature of air-water interface
e_a = actual vapor pressure of air, millibars

∇H_A = Average change in water-surface elevation adjusted for inflow, outflow, and rainfall, feet.

δ = net ground-water seepage, feet

Lakes with large inflows, outflows, and seepage rates usually produce large errors in water-budget computations; therefore, measurement of evaporation by energy-budget techniques may be substituted for the water budget to develop the mass transfer coefficient, N. Frenkiel[110] discussed the application of the combined energy budget-mass transfer method and concluded that the measurements required should extend for at least one year in order to maintain reasonable accuracy.

A study of the energy budget in Lake Ontario is described in an article by Rodgers and Anderson;[111] however, the data collected were from different periods, placing question on the validity of the results. Koberg[112] described methods to compute long-wave radiation from the atmosphere and reflected solar radiation from a water surface. The only parameters required were air temperature, air vapor pressure, and the ratio of measured solar radiation to clear sky solar radiation.

In an attempt to estimate the possible saving in water resulting from the use of lakes or existing reservoirs instead of cooling towers for the dissipation of heat, Harbeck[113] found that for the climatic conditions prevailing at that site in Texas for the dissipation of excess heat substantial savings in water often can be realized by using an existing reservoir or lake rather than cooling towers. Using data from the Lake Hefner study and a hypothetical lake, he applied the energy budget to the lake with excess heat added and made similar calculations for a cooling tower. Curves are presented showing the rise in temperature in the reservoir caused by the addition of excess heat and a comparison of reservoir evaporation and water used by evaporation. Another study conducted by Harbeck[114] on a lake used for cooling water purposes showed that the increase in evaporation caused by the power plant discharge, when expressed as a volume, is approximately 50% of the amount of heat added and is practically independent of reservoir contents. Evaporation was determined by the energy budget method and checked by a water budget.

The use of the energy budget method to predict temperatures in reservoirs for water quality control purposes is also discussed in papers by Raphael, [115] Delay and Seaders,[116] and Boyer.[117] In each case, conventional measurements of the parameters previously described were subjected to the usual calculation procedures, resulting in predictive models for temperatures in and downstream from the impoundments. A discussion of instrumentation for water temperature studies is presented by Moore[118] where it was concluded that temperature could be measured with an accuracy of 1°F.

A major problem encountered with respect to the prediction of temperature distributions in stratified reservoirs is the form of the temperature distribution.[119, 120]

Orlob proposed a mathematical model to simulate energy distribution in reservoirs and streams which appears to be a most significant contribution. The model, which utilizes the digital computer, was tested on Fontana Lake with good results when the complex hydrodynamics of the reservoir were considered. Orlob derived the following conclusions from his study:

1. The transfer of energy into deep stratified impoundments is accomplished by four primary mechanisms: advection, direct solar insolation, convective mixing associated with cooling at the surface, and "effective diffusion" identified with momentum transfer within the water body.

2. The transfer of energy from the epilimnion to the hypolimnion by diffusive mechanisms is limited at the thermocline by a strong density gradient.

3. A one-dimensional mathematical representation of energy transfer processes provides a satisfactory means of characterizing the thermal behavior of deep impoundments of the low volume-high discharge type.

4. Reservoir thermal behavior over annual or diurnal cycles can be simulated with a mathematical model, using as input information continuous records of hydrologic, meteorologic, and climatologic conditions of the system.

Using a statistical approach for the energy balance, and assuming the eddy conductivity,

Velz et al.[121] derived temperature distributions in a reservoir that appeared quite reasonable. Wunderlich and Elder[122] have proposed a simple graphical technique for determining temperature distribution in reservoirs that is quite useful for preliminary studies. The authors[123] utilized this method for predicting temperatures downstream from a proposed pumped storage project with apparent success.

It is of interest to note that Dake and Harleman[124] have successfully simulated the development of thermal stratification under laboratory conditions. Incoming radiation was assumed to be partially absorbed at the water surface, the remainder being absorbed exponentially beneath the surface. Heat was also assumed to be conducted downward by molecular diffusion. The artificial radiation source utilized was from mercury vapor and infrared lamps.

It should be noted that the real deficiency in any of these analytical models describing temperature distributions lies in the use of an assumed vertical diffusion coefficient. Dake and Harleman concluded that a general analytical approach should include the effect of internal radiation absorption. Neglect of mechanisms causing radiation absorption will lead to incorrect values of the diffusivity, as has been demonstrated in the literature.[125-127] Again quoting Dake and Harleman,[128] "There exists no satisfactory method of specifying a time and depth functional relation for the eddy diffusivity that could be used in a predictive mathematical model."

Heat Dissipation from Rivers

The computational methodology for the prediction of stream temperature in rivers differs somewhat from that used for reservoirs because of the difference in their physical characteristics. Because of the dynamic nature of a river, relatively short periods must be used to account for diurnal fluctuations in meteorological phenomena. Allowance must also be made for stream velocity and changes in hydraulic geometry.

There are two currently used methods for predicting stream temperatures. One method uses the energy budget and the other assumes the temperature decays downstream in an exponential manner.

Energy Budget Approach

Schroepfer[129] converted heat exchange quantities to incremental temperatures and applied the energy budget to the Mississippi and Minnesota rivers.

Application of this equation to several stretches of river proved to be reasonably successful when compared to observed data, both on naturally heated streams and those affected by a heated discharge.

Raphael[130] presented the following mathematical model, applicable to lakes and rivers

$$\frac{dt_w}{d\theta} = \frac{Q_t A + m_i (t_i - t_w)}{m_w} \qquad (20)$$

where

θ = time, hours
m_w = total mass of lake
t_w = mean temperature of lake, °F
m_i = inflow water mass
t_i = inflow water temperature, °F
A = lake surface area, acres
Q_t = total surface heat transfer per unit of time, BTU ft^{-2} hr^{-1}

Substitution of volumes for mass and solution for given increments of time yields a simplified and workable equation. For application to river work, the outflow temperature of a reach is assumed to be the inflow temperature of the subsequent downstream reach. The article exemplifies the method by using a small reservoir and a river.

Messinger[131] used the energy budget approach to predict the temperature profile of a heated stream in the Susquehanna River and found large discrepancies in the results. A conventional energy budget approach was utilized; however, he attributed the errors to inadequacies in the measurement of solar and atmospheric radiation. Shading of the river by trees and hills was assumed to be the major source of error.

Garrison and Elder[132] developed a water temperature prediction model based on fundamental laws of physics and verified its use on two sets of field data. The complete conservation of energy balance utilized the following equation

$$u\rho cT\Delta y\Delta z - \left[u\rho cT + \frac{\partial(u\rho cT)}{\partial x}\,\Delta x \right]\Delta y\Delta z + v\rho cT\Delta x\Delta z - \left[v\rho cT + \frac{\partial(v\rho cT)}{\partial y}\,\Delta y \right]\Delta x\Delta z$$

$$+ w\rho cT\Delta x\Delta y - \left[w\rho cT + \frac{\partial(w\rho cT)}{\partial z}\,\Delta z \right]\Delta x\Delta y + \Sigma Q\Delta x\Delta z = \frac{\partial}{\partial t}(\rho cT\Delta x\Delta y\Delta z) \qquad (21)$$

where

$$\Sigma Q = Q_s - Q_{rs} + Q_a - Q_{ra} - Q_b - Q_e$$
$$- Q_L - Q_h + Q_{ad} - Q_c$$

u, v, w = velocity components in x,y,z directions

$\frac{\partial}{\partial x}, \frac{\partial}{\partial y}, \frac{\partial}{\partial z}$ = rates of change in x,y,z directions

ρ = mass density of water, lb sec²/ft⁴

T = mean temperature of water in control volume, °F

Q_s = rate of heat flow into control volume from solar radiation, BTU/ft²/day

Q_{rs} = rate of heat flow out of control volume: reflected solar radiation, BTU/ft²/day

Q_a = rate of heat flow into control volume: atmospheric radiation, BTU/ft²/day

Q_{ra} = rate of heat flow out of control volume: reflected atmospheric radiation, BTU/ft²/day

Q_b = rate of heat flow out of control volume: back radiation, BTU/ft²/day

Q_e = rate of heat flow out of control volume: direct evaporation, BTU/ft²/day

Q_L = rate of heat flow out of control volume: latent heat loss, BTU/ft²/day

Q_h = rate of heat flow out of or into control volume: conduction, BTU/ft²/day

Q_{ad} = rate of heat flow in or out of control volume: rainfall or tributary flow, BTU/ft²/day

Q_c = rate of heat flow in or out of control volume: bed conduction, BTU/ft²/day

y_o = mean flow depth based on surface width

t = time, day

c = specific heat of water, BTU/lb/°F

Introduction of turbulent diffusion concepts, using short stream lengths, 24-hour time averages, and steady-state flow conditions, reduced equation 21 to:

$$\frac{T}{T_o} = \frac{t}{T_o U}\,\frac{\Sigma Q}{c\rho y_o} + 1 \qquad (22)$$

which states that the space-average dimensionless water temperature over a 24-hour cycle of any distance x from a reference station, x = o, in which $T = T_o$, is directly proportional to the product of ΣQ and x and inversely proportional to the product of T_o, U, c, ρ, and y_o.

The results of two field measurements showed excellent agreement of the observed temperature values and the proposed theory, which applies to the well mixed conditions existing at the test sites. This model appears to have great potential.

Exponential Decay of Transient Temperatures

LeBosquet[133] proposed a mathematical basis for predicting heat loss in a flowing stream using river flow, temperature differential between water and air, hydraulic characteristics, and a heat loss coefficient. While the method appears to be quite useful, it is somewhat limited because the value of the heat loss coefficient must be obtained from field observations. LeBosquet found the coefficient to vary from 6 to 18 BTU/ft²/hr/°F of excess water temperature over air.

The mathematical model proposed is

$$\frac{dF}{dt} = -\frac{KA_sF}{L} \qquad (23)$$

where

K = heat loss coefficient, BTU/ft²/hr/°F of excess temperature of water over air

F = excess temperature of water over air at distance, D miles, °F

A_s = surface area, ft²

L = weight of water, lb

40

Integration and simplification of this equation yields

$$\frac{Q \log_{10} \dfrac{F_A}{F}}{0.0102 \, WD} = K \qquad (24)$$

where

Q = average discharge, cfs
F_A = initial excess temperature, °F (water over air)
W = average stream width, ft
D = reach distance, miles.

It should be noted that the driving force utilized by LeBosquet is incorrect, inasmuch as the water will tend to approach some equilibrium water temperature rather than the temperature of the air, and the two may not be equal in the short run though they will in the long run for natural conditions.

Gameson et al.[134] made an exhaustive study on heated discharges into the Thames Estuary and proposed the following mathematical model describing the process

$$\frac{d\theta}{dt} = -\frac{f}{z}\,\theta \qquad (25)$$

where

θ = initial temperature increment, °C
f = exchange coefficient, cm/hr
z = average river depth, cm
t = time, hr.

Using statistical methods, they defined the natural or equilibrium temperature, T_E, as

$$T_E = 0.5 + 1.109 \, T_a \qquad (26)$$

where

T_a = air temperature, °C.

The solution to the equation utilized was

$$\theta_t = \theta_o \exp\left[-_o\!\int^t \frac{f}{z}\,dt \right] \qquad (27)$$

where

θ_o = initial excess temperature,
θ_t = excess temperature after a time length equal to two tides.

Values of the exchange coefficient for the Thames Estuary were found to average approximately 4 cm/hr. Gameson et al.[135] utilized the same model on the River Lea, which clearly demonstrated the profound effect of the heated discharge on the thermal regime of the river. The exchange coefficient in this case was found to vary from 1.66 to 3.83 cm/hr. This study concluded that the exchange coefficient was unrelated to wind velocity, which is obviously erroneous.

It is interesting to note that the model proposed is similar to that utilized by English workers to describe reaeration and thus demonstrates the effect of turbulence on heat dissipation.

Velz and Gannon[136] assumed that the rate of temperature change is directly proportional to the rate of heat loss from the water surface, or

$$\frac{dT_w}{dt} = -\frac{H}{62.4b} \qquad (28)$$

They presented the following solution to express the rate of temperature change

$$A = -224,640 \int_{T_1}^{T_2} \frac{dT_w}{a[V_w - V_E + \beta(T_w - E)]} \qquad (29)$$

where

A = surface area of the river between the points where $T_w = T_1$ and $T_w = T_2$, ft²/cfs stream flow
V_w = saturation vapor pressure corresponding to surface temperature T_w, in. Hg
V_E = saturation vapor pressure corresponding to surface temperature E, in. Hg
E = equilibrium temperature, °F
α = 0.00722 H_v C(1 + 0.1 W)
T_w = water temperature, °F
t = time, hr
H = rate of heat loss from the water surface, BTU ft⁻²hr⁻¹
b = depth of the river, ft
62.4 = product of the water density and heat

capacity; BTU ft^{-3} °F^{-1}

β = $(1.8 + 0.16\text{ W})$

H_v = latent heat of vaporization, BTU lb^{-1}; at assumed water temperature

C = evaporation coefficient ranging from 10-15

W = wind speed at 25 feet above water surface, mph

T_a = air temperature, °F

V_a = vapor pressure of the air, in. Hg

H_s = heat gain by solar radiation; BTU ft^{-2}hr^{-1} (measured).

They derived a relationship for the long-term unheated equilibrium water temperature as follows

$$(1.8 + 0.16\text{W}) \text{ E} + 0.00722 \text{ H}_v\text{C} (1 + 0.1\text{W}) \text{ V}_E = \quad (30)$$
$$(1.8 + 0.16\text{W}) \text{ T} + 0.00722 \text{ H}_v\text{C} (1 + 0.1\text{W}) \text{ V}_a + \text{H}_s$$

This equation is solved by successive approximation, assuming values for the equilibrium temperature and using Meyer's evaporation formula. The resulting value of E, the equilibrium water temperature, is then used in the following working equation to estimate the required water surface area for cooling

$$A = -224{,}640 \sum_{T_1}^{T_2} \frac{\Delta T_w}{\alpha(V_w - V_E) + \beta(T_w - E)} \quad (31)$$

The total increment of temperature between the initial heated condition and the desired downstream temperature is divided into equal increments (ΔT_W) with T_W and V_W the mean temperature and mean saturation vapor pressure in each increment. Long-term weather averages are used in the computations.

Duttweiler[137] proposed a mathematical model for stream temperature by equating the heat gained in an incremental reach of stream to the time change in enthalpy of the water in the reach. As in previous studies, he assumed that the time rate of temperature increase is proportional to the deficit between the actual temperature and some equilibrium temperature. The resulting one dimensional model was

$$\frac{\partial T}{\partial t} + v\frac{\partial T}{\partial x} = \frac{1}{\rho c} \frac{\lambda}{z} (T_E - T) \quad (32)$$

where

T = water temperature, °F

t = time, hr

v = velocity at x and t, cm/hr

x = distance along the stream, cm

ρ = water density, g/cm^3

c = heat capacity of water

λ = parameter dependent upon atmospheric conditions

z = hydraulic depth of the stream or the cross-sectional area divided by the surface width

T_E = equilibrium water temperature, °F

Using Lagrangian coordinates, the expression for river temperature becomes

$$\frac{dT}{dt} = \frac{1}{\rho c} \frac{\lambda}{z} (T_E - T) \quad (33)$$

and its position may be found from

$$X - X_o = \int_{t_o}^{t} v dt \quad (34)$$

It is obvious that $\lambda (T_E - T)$ is the net heat transfer rate through the air-water interface if evaporation losses are excluded, or

$$\lambda (T_E - T) = \Delta H + H_c \quad (35)$$

Duttweiler presents values for λ based on published meteorological data and gives solutions to equations 32 and 34 in integral form.

Edinger and Geyer[138] proposed that the net rate of heat transfer through the air-water interface was represented by

$$\Delta H = -K(T_s - E) \quad (36)$$

where

ΔH = net heat transfer rate through air-water interface, BTU/ft^2/day

E = equilibrium temperature, °F

T_s = water surface temperature, °F

K = exchange coefficient, BTU/ft^{2}day/°F

Nomograms for determining E when K is known were presented and methodology defining the exchange coefficient was given.

Assuming that the river reach considered was at steady state, they proposed one-dimensional and two-dimensional models for temperature prediction. The one dimensional model was

$$\rho C_P U d \frac{\delta T}{\delta x_1} = -K(T - E) \qquad (37)$$

and the two-dimensional model, which ac-

counts for lateral mixing and advection, was

$$U \frac{\delta T}{\delta x_1} + D_2 \frac{\delta^2 T}{\delta x_2^2} + \frac{K}{\rho C_P d}(T - E) = 0 \qquad (38)$$

The solution presented for equation 37 was

$$\frac{T - E}{T_m - E} = \exp\left[-\frac{K x_1}{\rho C_p U d}\right] \qquad (39)$$

and for equation 38

$$\frac{T - E}{T_o - E} = \exp\left[-\frac{C_2 + \alpha^2}{C_1} x_1\right] \exp\left[-\alpha x_2\right] \cdot \frac{\exp\left[-2\alpha(W - x_2)\right]}{1 + \exp\left[-2\alpha W\right]} \qquad (40)$$

where

ρ = water density, lb/ft³
C_P = heat capacity of the water, BTU/lb
U = mean velocity of the stream at x_1, ft/day
d = mean depth of the stream at x_1, ft
$\frac{\delta T}{\delta x}$ = longitudinal temperature gradient, °F/ft
T = water temperature, °F
x_1 = longitudinal distance on the stream, ft
D_2 = lateral mixing coefficient, assumed constant over the channel, ft²day⁻¹
X_2 = distance across the stream, ft
T_o = temperature of a thermal discharge at the point $x_1 = x_2 = 0$, °F
C_1 = U/D_2, ft⁻¹
C_2 = $\dfrac{K}{C_P d D_2}$, ft⁻²
α = coefficient depending on boundary conditions
W = stream width, ft
T_m = water temperature at upstream end of reach, °F

Edinger and Geyer's study includes basic concepts of the energy balance, applied aspects with respect to reservoirs and rivers, and operational methods. Heated water investigations are discussed, including a discussion on instrumentation and the mechanisms of heat exchange with the atmosphere. Several useful models of temperature distribution in reservoirs and rivers are presented and ensuing applications of the methods presented demonstrate the methodol-

ogy. However, the difficulties in applying the nomographs are not sufficiently stressed. The systems studied are primarily steady state systems, or the temperatures are known and transfer coefficients computed rather than the usual case of attempting to predict the temperatures.

Examination of the non-English literature demonstrates approaches similar to those of LeBosquet, where the temperature downstream from a power plant effluent is assumed to decay exponentially and "river constants" are derived from field measurements. The real problem with this type approach, of course, is the validity of the assumptions; i.e., steady state conditions, non-varying river flows, vertically and laterally uniform temperature distributions, and the requirement that the equations apply only downstream of the river section where complete mixing is achieved. Probably the most notable of these studies are Halleaux,[139] Goubet,[140, 141] and Mandelbrot.[142] Further work at Electricité de France[143, 144] shows that the energy budget methods can be used in predicting river temperatures.

The work of the advanced seminar at Johns Hopkins University,[145] where the following conclusions were drawn, should also be mentioned:

1. The quantitative determination of heat transfer within a natural body of water, or between a body of water and its surroundings, is extremely difficult and complex.

2. Temperature change within a body of water is primarily effected by the operation, either independently or jointly, of two mecha-

nisms: turbulent mixing of two or more batches of water of different temperatures eventually resulting in a single batch of water at the weighted mean temperature; and heat exchange of the water with its surroundings, primarily the atmosphere, governed by the mechanisms of conduction, convection, evaporation, condensation, and radiation.

3. Temperature change is affected to only a minor degree by molecular diffusion and conduction within the body of water.

4. Forecasting of heat loss from artificially heated batches of water may be attempted by use of certain of three principal techniques depending on the availability of data and on the adherence of the specific situation to certain specialized requirements.

5. Where heat is discharged to a stream from a point source, and where complete vertical mixing may be assumed, turbulent mixing may be considered the dominant heat transfer mechanism until horizontal mixing is complete; that is, until that downstream transect is reached where the cross-section temperature is uniform from bank to bank.

6. Assuming that there is no horizontal temperature gradient, either as a result of effective turbulent mixing as described in No. 5 above, or as a result of heat discharge from a thoroughly diffused source, and assuming complete vertical mixing, heat loss may be evaluated by either use of the heat budget method or equilibrium temperature method.

7. The complexity of the heat budget method makes this method unfeasible for field use except where extremely precise, reliable, and rugged instrumentation is available.

8. At present, not all of the instrumentation required by No. 7 above is available.

9. The relative simplicity of the equilibrium method, coupled with its feasibility for use under field conditions, suggests its use for forecasting heat loss both in specific field situations and in generalized design situations using combinations of ambient conditions selected according to statistical considerations.

Jaske[146,147] presents computer programs for determining the temperature distribution in the Columbia River. Variation in hydraulic characteristics of the river was accounted for by dividing the river into a central channel and two side channels and applying an energy budget similar to Raphael's.

Dingman et el.[148] presented a method for calculating the length of ice-free river that develops below a power plant during the winter. Their method uses daily averages of meteorologic variables and "by use of a computer program allows the use of several of the many previously developed empirical and semi-empirical formulae for the heat budget terms."

Finally, the work of Roesner[149] should be mentioned because of the somewhat unique application of the surface renewal theory as proposed by Danckwerts[150] to account for the effects of turbulence on heat transfer across the interface. Roesner developed two models to describe the net rate of heat transfer through the interface of a flowing stream. The only difference in the models was the depth of the fluid elements utilized in the surface renewal scheme. One model assumed that the elements were infinitely deep and the other assumed that the depth was limited to the thickness of the film. Application of the models to field data was favorable and the author concluded that, "While turbulence does significantly affect the net rate of heat transfer through the air-water interface, it probably does not significantly affect short-term temperature prediction. However, neglecting to account for the turbulence level introduces a bias in the time rate of temperature change so that long-term averages in the order of a week or month may be significantly affected during periods of warming or cooling."

Equilibrium Temperature

If a body of water at a given initial temperature is exposed to a given set of constant meteorological conditions, it will tend to approach another particular temperature asymptotically. It may warm by gaining heat or cool by losing heat. Theoretically, after an infinite period of time the temperature will become constant, and the net heat transfer will be zero. This final temperature has been called the equilibrium temperature, E. At equilibrium, the heat gained by absorbing solar radiation and long-wave radiation from the atmosphere will exactly balance the heat lost by back radiation, evaporation, and conduction.

That is

$$H_s + H_a + H_b + H_e + H_c = 0 \qquad (41)$$

In actuality, the water temperature rarely equals the equilibrium temperature because the equilibrium temperature itself is constantly changing as the local meteorological conditions change. The equilibrium temperature will generally rise during the day when solar radiation is greatest, and fall to a minimum at night when solar radiation is absent.

A daily average equilibrium temperature may be computed using daily average values of radiation, temperature, wind speed, vapor pressure, etc., and this daily average will reach a maximum in midsummer and a minimum in midwinter. Other time periods, such as monthly, may also be used for averaging. Since the actual water temperature always tends to approach, but lags behind the equilibrium temperature, it will usually be less than equilibrium in the spring when temperatures are rising, and greater than equilibrium in the fall when temperatures are falling. During a one-day period the equilibrium temperature usually varies from above the actual water temperature during the day to below the actual water temperature at night, forcing the water temperature to also assume a similar cycle.

The annual variation of temperature has been fitted by a sine curve by Ward,[151] and Duttweiler[152] presents a review of attempts to utilize a sine function to fit daily and yearly temperature fluctuations. As stated by Duttweiler, "It is not suggested that the diurnal variation, for example, be considered in analyzing stream conditions over several months, nor that the annual variation be considered in analyzing stream conditions over a few hours. However, the presence of these cycles should be kept in mind, especially when dealing with stream characteristics which are particularly sensitive to temperature change."

Roesner[153] showed that though the models commonly used appear quite different in the form of their solutions, most of them are quite similar and can be expected to produce nearly the same answers when applied to a given problem. It was shown that Schroepfer's model and the Advanced Seminar model differ only in

1) the method of computing long-wave at-mospheric radiation

2) the units and empirical coefficients used, and

3) the inclusion of heat exchange by eddy diffusion by the Advanced Seminar.

It was also shown that the differential forms of the models of Gameson, Gibbs, and Barnett, Velz and Gannon, Duttweiler, and Edinger and Geyer are very similar. For easy reference, the equations are tabulated in Table 7 (see page 58). Consider a river reach which is receiving a thermal discharge and note:

TABLE 8

EXPERIMENTAL DATA FOR THREE-DIMENSIONAL JETS

AUTHOR	TYPE OF FLUID	ρ_s/ρ_o	L/d_o	C_2
Albertson	Air	1	250	0.081
Keagy	Mixed gases	1.03	50	0.086
Becher	Air	1	35	0.077
Forstall	Water	1.01	30	0.078
Taylor	Air	1	30	0.076
Corrsin	Warm Air	1	25	0.076
Hinze	Warm gases	1	20	0.078
Folsom	Air	1		0.095
Poreh	Water	1	10	0.065

---

1. If mean daily or monthly temperatures are used for the natural stream temperature, the equilibrium temperature of Duttweiler (T_E) and of Edinger and Geyer (E) can be considered equal to the natural river temperature. Thus ($T - T_E$) and ($T - E$) can be equated to Gameson, Gibbs, and Barnett's excess temperature θ, i.e.,

$$\theta = (T - T_E) = (T - E).$$

2. The time of travel, t, and the longitudinal distance along the stream, x, are related as

$$\frac{dx}{dt} = U$$

where U is the mean velocity of the stream at the point x; thus

$$\frac{dT}{dt} = U \frac{dT}{dx}$$

45

3. If the time base for travel time is short compared to the time interval for which the natural stream temperature is averaged, then E and T_E can be considered constant and

$$\frac{d\theta}{dt} = \frac{d(T - T_E)}{dt} = \frac{dT}{dt}$$

and

$$\frac{d\theta}{dt} = \frac{d(T - E)}{dt} = \frac{dT}{dt} = U\frac{dt}{dx}$$

4. By use of items (1) and (3) above, it is observed that

$$f = \frac{\lambda}{\rho C} = \frac{K}{\rho C}$$

The essential difference is that λ and K are defined in terms of atmospheric variables and parameters, while f is strictly empirical.

5. It is observed that the net rate of heat loss from the river H, in Velz and Gannon's model, can be equated as

$$H = \lambda(T - T_E) = K(T - E).$$

Roesner concluded, "Thus for the conditions stated in item 1 the basic models of Gameson, Gibbs, and Barnett; Duttweiler; Edinger and Geyer; and Velz and Gannon can be considered identical. The solutions, however, will not give identical answers because of the different approaches taken in integrating the individual models. The other models reviewed here are unique either in their derivation or in the methods used to obtain the working equation and thus cannot be compared.

"The models described above define the state of development of temperature prediction equation for streams. Note that all the models presented here are one-dimensional except for Edinger and Geyers' two-dimensional model which is quite simplified."

Remote Sensing of Thermal Pollution

Though this review does not concern itself with instrumentation, it was thought advisable to briefly note the work already done on the use of remote sensing techniques to determine the discharge and spread of heated waters. Perhaps most outstanding has been the use of the NIMBUS satellite high resolution infrared imagery to detect currents, upswellings, and temperature contours of the Great Lakes.[154] Though some of the needed information in this area is still classified, sufficient data are known to enable one to use the technique and its airplane and laboratory versions to study the patterns of temperature distribution on water surfaces. Present capabilities in infrared imagery (not photography) are such that a number of commercial firms are available on a contract basis to provide complete service. Infrared wave lengths, in the 1.5 to 14 micron range, are used in hydrologic studies, because in this region the infrared energy emitted by land and water bodies is a maximum and atmospheric absorption is at a minimum.[155] A typical scanner is sensitive to $\frac{1}{4}°$ C surface temperature changes and can be calibrated with surface instrumentation to $\frac{1}{2}°$ C. Infrared radiometers sensitive to $0.01°$ C are not uncommon, but obtain data at such a low rate to be useless for large-scale studies.[156]

A brief evaluation of the use of remote sensors in hydrology was presented at the International Conference on Water for Peace.[157] At that time, photographic sensors were still most favored because of the easy interpretability and the minimum data reduction and specialized training needed for their use. Infrared sensors offer hope of great things in thermal pollution work, but are hampered in a quantitative—not qualitative— sense at present by problems of emissivity measurement. In any case, infrared imagery is only a function of the surface temperature of the waters and does not give any indication of the thermal profile in depth.

An earlier paper provided in detail a large amount of data on patterns of thermal pollution, by using aerial photography and including a number of pictures of power plant outfalls.[158]

MECHANISM OF HEATED WATER DISCHARGES

There are two basic methods of discharging heated waters. The heated effluent can be discharged as a layer (subsurface or surface) or as a mixing jet. The former results in a small volume with the same temperature as the effluent (if no mixing whatsoever occurs). The latter, mixing jet, is designed to minimize the temperature rise at any point by mixing with a large part of the total volume of available water. At the end of the zone in which this mixing occurs, the final mixed temperature throughout the river cross-section should be

$$T_M = T_R + \frac{Q_P}{Q_R} \Delta T \qquad (42)$$

where

T_M = mixed temperature

$\Delta T = T_P - T_R$

T_P = temperature of plant effluent

T_R = temperature of ambient fluid

Q_P = plant flow

Q_R = river flow.

The layered discharge can be obtained by use of a canal, frequently used, for example, by plants in the TVA system. Pipes can be used also, although for large volumes of cooling water, unless sufficient flow area were provided, significant mixing would be expected to occur. Such pipes could be at or below the water surface. The Widows Creek Steam Plant uses a series of rectangular conduits emptying into a small harbor. The condenser water may also be discharged into another water body and hence into the main receiving body. The effluent from the New Johnsonville Steam Plant (in Tennessee) empties into a coal barge harbor and thence into the main river.

Discharge for complete mixing is usually accomplished by use of a multi-port diffuser placed on the river bottom perpendicular to the flow. Harleman, Hall, and Curtis[159] performed extensive model studies to form a basis for design of a diffuser system for the proposed Browns Ferry Nuclear Power Plant. Studies were made to determine the following: distance downstream to uniformly mixed profile, effect of jet port diameter and spacing, effect of jet angle, effect of differing steady river discharges, scour produced by the jets, upstream extension of a thermal wedge, and effect of unsteady river flows on the diffuser system characteristics. Harleman[160] discusses these problems further.

If the cooling water is discharged as a layer, a portion will be carried downstream. A portion may also move upstream as a thermal wedge.

If the mixing jet is used for discharge, an upstream thermal wedge may also form, being again subject to surface heat exchange. The downstream layer will mix laterally (and perhaps vertically) with the ambient fluid as a result of initial differences in velocity and momentum. Upon dissipation of these differences, dispersion of the heated mass will continue as a function of the processes of turbulent diffusion and mixing. Superimposed on both the wedge and downstream processes is the influence of heat exchange between water and the atmosphere. This heat exchange is the only consideration downstream from the fully mixed section.

Vertical mixing does occur in many layer discharges, though most are designed to minimize such mixing. However, if the temperature of such a layer would be too great, or too long a wedge would form, etc., the designer could choose to attempt to cause partial mixing with a resultant layer at a lower temperature. While not as rapid for heat dissipation to the atmosphere, some such benefits are retained while possibly lessening some threats to water quality.

It should be noted that there are limits to the discharge of waste heat to a water body. It has been observed[161] that, "Such techniques cannot be expected to solve the problems associated with the discharge of large heat loads into moderate volumes of receiving water. Any heated discharge which will promote excessive temperature rises will require the use of cooling devices." The limitation should perhaps be rephrased to include any situation created by heated discharges which would reduce any water quality parameter below acceptable levels.

Formation and Description of Warm-Water Wedges

If a heated effluent is discharged into a body of water, several situations may develop depending on the characteristics of the receiving fluid and the discharge. Issuing with little momentum, the discharge may simply float onto the water surface; with higher momentum, the discharge (or jet) mixes with the ambient fluid to the extent dictated by the original momentum to be dissipated. If complete mixing does not occur and density gradients exist, then the flow characteristics of the receiving water are important. The heated water may be swept away downstream, or a wedge may be formed upstream. Treatment of the mixing problem and sweep-away conditions are noted under the section on momentum jets. In this section, we shall treat the problem of the heated wedge extending upstream above the underlying colder water. Some mention of subsurface layers will also be made. In addition to the possible direct effect of higher temperatures on aquatic organisms, there is the possibility of recirculation of heated water through the intake channel. This not only decreases steam plant efficiency, it could lead to compounding the temperature problem. It is, therefore, important to have means for evaluating lengths, depths, etc., for these wedges. There do exist means for withdrawing colder water, such as by withdrawal beneath skimmer walls.[162-164] However, even here it is necessary to know the length and depth of the wedge to predict how much warmer water might flow beneath the wall.

Problem Formulation — Dimensional Considerations

The wedge problem requires solution of the equations of stratified flow. Ordinary open channel flow is actually a stratified flow, with the water lying beneath a lighter fluid, the air. The heated discharge problem deals (as it has been thus far treated) with a two-layered system with hotter, lighter water above the colder water. Another application of stratified flow principles would attack the problem of a heated layer lying beneath the water surface. This (called an interflow) could occur when solar heating raises the receiving water surface temperatures to levels above that of the heated discharges. Treatment of such flows will not be discussed here in detail. At the present, surface layers are of greater importance and more frequent occurrence.

Dimensional analysis indicates that the length, as shown in Figure 11 (see page 49), can be expressed as a function of the total depth H as:

$$\frac{L}{H} = f\left(F_\Delta, Re_\Delta, \frac{H}{B}, V_r\right) \qquad (43)$$

where

F_Δ = Upstream densimetric Froude number = $\dfrac{u_o}{\sqrt{\dfrac{\Delta\rho\ g\ H}{\rho_2}}}$

Re_Δ = Upstream densimetric Reynold's number = $\dfrac{\sqrt{\dfrac{\Delta\rho}{\rho}\ g\ H}\ \ H}{\upsilon}$

V_r = u_{jet} / u_o

B = width of the channel

It has been observed by Chikwendu and Francis,[165] and supported by other data, that the velocity ratio, V_r, only has effect on the degree of initial mixing and has no apparent influence on wedge length or shape upstream from the outlet. This quantity can, therefore, be dropped from consideration.

Of special importance in this problem is the densimetric Froude number. Theoretically, just as in ordinary free surface flow, a Froude number greater than one implies that a disturbance cannot propagate upstream, since the stream velocity is greater than the velocity of propagation. Harleman,[166] Bata,[167] TVA,[168] and others note this theoretical value. Note that this Froude number is based on the flow in the river between the intake and outlet, as a certain portion of the approaching river flow is diverted through the intake. Harleman[169] states that in practice $F_\Delta > 0.7$ is sufficient to prevent a wedge. Keulegan,[170] in studies on salt wedges, found that the critical value of F_Δ was 0.75. TVA used this value in studies for the Browns Ferry Nuclear Plant.[171] Chikwendu and Francis,[172] studying wedges in a curved channel, extrapolated to zero wedge length at $F_\Delta = 0.83 \pm 0.2$. This latter value might have been expected to be somewhat lower, since wedge lengths for the curved channel were significantly less than those in straight channels. It can be seen that wedges should be slight or non-existent for Froude numbers of 0.7-0.8 or greater.

Basic Equations

Harleman[173] noted that the equation for the

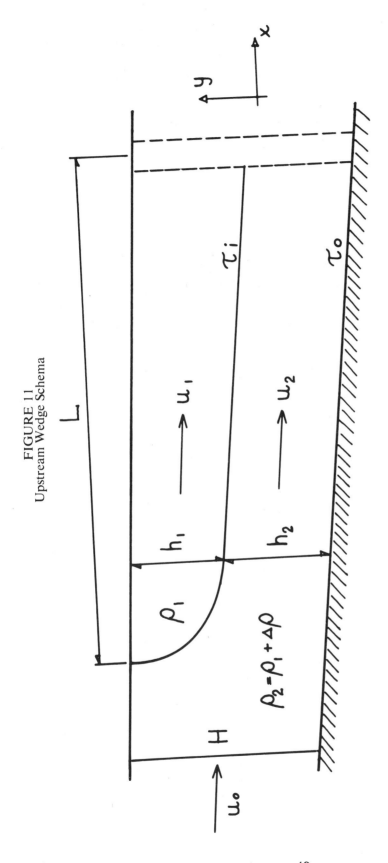

FIGURE 11
Upstream Wedge Schema

49

slope of the interface between the wedge and the flowing water could be written as:

$$\frac{dh_2}{dx} = \frac{\dfrac{f}{8gh_2}\, u_2\,|\,u_2\,| - \dfrac{f_i}{8gh_2}\left(\dfrac{H}{H\text{-}h_2}\right)|\,u_1 - u_2\,|\,(u_1 - u_2)}{\dfrac{\Delta\rho}{\rho}\left[\,F_2{}^2 + F_1{}^2 - 1\,\right]} \tag{44}$$

where

$F_1 = u_1 \,/\, \sqrt{g(\Delta\rho/\rho)h_1}$
$F_2 = u_2 \,/\, \sqrt{g(\Delta\rho/\rho)h_2}$
$h_1 = $ depth, layer 1
$h_2 = $ depth, layer 2
$H\; = $ total depth
$f\;\; = $ friction factor, water-bottom interface
$f_i\;\; = $ friction factor, interface between layers
$u_1 = $ velocity, layer 1
$u_2 = $ velocity, layer 2
$\Delta p = $ density difference $= \rho_2 - \rho_1$
$p\;\; = (\rho_1 + \rho_2)\,/\,2$

In writing the equation, vertical accelerations and side wall shear effects are neglected, and only mean velocities are considered. The ordinary free surface Froude number, based on total depth and average velocity, is assumed much less than unity. Therefore, the total depth can be assumed constant relative to changes in interfacial position, or $H = h_1 + h_2 = $ constant.

Other assumptions[174] are

1. One-dimensional flow in each layer. Harleman notes that the hydraulic radius may be adjusted in computing the friction factor, affording some improvement for two and three dimensional flows.

2. Uniform density in each layer. This essentially states that there is no interfacial mixing between layers. Since interfacial stability represents a balance between inertial, gravitational, and viscous forces, both the Reynolds number and a densimetric Froude number are important. Keulegan[175] has proposed a stability parameter, θ, defined as

$$\theta = \frac{1}{(F_2)^2 \, Re_2} \tag{45}$$

where the subscripts represent values calculated based on depth and flow in layer 2 and F is the densimetric Froude number with Re the Reynolds number. Critical values have been defined by data from Ippen and Harleman[176] and Keulegan to be, as reported by Harleman[177]

Laminar: $\quad \theta_c = \dfrac{1}{Re_2}$

Turbulent: $\theta_c = 0.18$

For $\theta > \theta_c$, mixing should not occur. The criteria amount to specifying $F_2 > 1.0$ for instability in the laminar case. Harleman and Stolzenbach discuss work in turbulent flows which show similar trends of little or no mixing for $F_2 > 1$. While a sharp interface is never obtained, a transition zone always exists. Harleman and Stolzenbach[178] state, "An analysis of correction factors for the uniform density assumption shows that for even moderately wide transition zones, the error introduced is small." Bata[179] wrote Equation 44 in terms of the length and depth of the wedge.

$$\frac{-\dfrac{1}{F_o{}^2}\,\eta^3\,(1-\eta)^3 - (\eta-1)^3 + C_Q^2\eta^3}{\alpha\,(1 - \eta + C_Q\eta)^2 + (1-\eta)^3}\, d\eta = \frac{f}{8}\, d\left(\frac{x}{h}\right) \tag{46}$$

in which $\eta = h_2/h$, $\alpha = f_i/f$, $C_Q = q_1/q_2$, and

$$F_o = \frac{u_o}{\sqrt{\dfrac{\Delta\rho}{\rho}gh}}$$

where $g_1, g_2 = $ discharge in layer 1, layer 2.

Other terms as previously defined.

Given appropriate boundary conditions, Equation 46 can be solved for a particular case. At the place where the wedge begins, there is a critical depth, and the interface changes rapidly, or

50

$$\frac{dh_2}{dx} \rightarrow \infty \qquad (47)$$

Since the numerator must equal zero, from Equation 46 one can write:

$$-\frac{1}{F_o^2} \eta^3 (1 - \eta)^3 - (\eta - 1)^3 + C_o^3 \eta^3 = 0 \qquad (48)$$

The foregoing equations require knowledge of friction factors: f, between the bottom layer and the stream bed, and f_i, between the two layers. While such factors are not yet well defined, work by several investigators implies similarity to ordinary open channel relationships. The factor f is determined as usual. A means of evaluating f_i was used by TVA.[180] The Reynolds number of the lower layer, $Re_2 = \dfrac{4U_2 h_2}{v}$ is used with a pipe friction diagram, assuming a hydraulically smooth interface.

Harleman and Stolzenbach[181] suggest factors β_1 and β_2 to modify the values of the hydraulic radius of the two layers to account for two- and three-dimensional effects, such that

$$Re_1 = \beta_1 h_1$$

$$Re_2 = \beta_2 h_2$$

$$B_1 = \frac{1}{1 + 2h_1 / B} \qquad (49)$$

$$B_2 = \frac{1}{2 + 2h_2 / B}$$

$$B = \text{channel width}$$

It can be seen that this corresponds to assuming free surface flow for the upper layer and a bottom flow which is enclosed on all sides. It should be noted that this may give greater importance than justified to the interfacial friction, but Ippeh and Harleman[182] have provided data indicating less than 10% error in f by this means.

Solution of the Upstream Warm-Water Wedge Equation

When the upper layer is stagnant ($u_1 = 0$), the basic equation can be integrated directly[182] yielding for the shape of the wedge:

$$f\frac{x}{h} = \frac{2}{F_o^2} (\eta^4 - F_o^{8/3}) + \frac{1}{F_o^2} \frac{8}{3} \alpha (\eta^3 - F_o^{6/3}) +$$
$$\frac{4\alpha (1 + \alpha)}{F_o^2} (\eta^2 - F_o^{4/3}) + \frac{8}{F_o^2}[\alpha (1 + \alpha)^2 - F_o^2] (\eta - F_o^{2/3}) -$$
$$\frac{8\alpha}{F_o^2}[(1 + \alpha)^3 - F_o^2] [\ln (1 + \alpha - F_o^{2/3}) - \ln (1 + \alpha - \eta)] \qquad (50)$$

Terms as previously defined.

Setting $\eta = 1$ allows one to evaluate L, the total length of the warm water wedge. A zero wedge length is predicted for $F_o = 1$. However, field results indicate F_o equal to about 0.75 is the critical value. Therefore, TVA, in its Browns Ferry studies[183] adjusted the equation to yield zero wedge length at $F_o = 0.75$.

Wedge into Discharge Channel

The existence of a cold water wedge upstream into the discharge canal is a means of minimizing mixing of the condenser water with the receiving fluid.[184] For the cold water wedge, the average velocity, u_2, of the lower layer is zero, and,

therefore, $F_2 = 0$. This is similar to an arrested saline wedge.

The basic Equation 44 reveals that F_1, the Froude number in the upper layer in the canal, is less than 1, and continuity shows that F_o for the canal is less than 1. Harleman[185] observed that the heated discharge undergoes a rapid expansion at the junction as it spreads laterally and longitudinally in the river. Hence, a critical depth is established in the upper layer at this point, for $F_{1B} = 1$. This criterion enables evaluation of the upper layer depth at the junction to be

$$h_1 = h_o F_o^{2/3} \qquad (51)$$

Presumably if little mixing occurred, the depth

51

of the heated layer in the main channel would be the same as that at the mouth of the outlet.

As in the river, F_o controls the formation of a wedge. Theoretically, for F_o greater than 1, no wedge forms, though this criterion may be subject to revision as before. If it is desired to inhibit mixing, a proper balance must be struck. Too high a Froude number will yield too short a wedge to fully prevent mixing, and too low a value of F_o will make necessary extremely long discharge canals. More work is needed to define appropriate values for such designs.

Wedges Extending to Intake

In some instances, the upstream warm wedge could reach upstream to the intake and beyond. To protect against recirculation of this heated water, skimmer walls are frequently used to cause the colder, bottom waters to be withdrawn. While such a wall does cause a drawdown of the interface in the vicinity of the junction, the effect of this on the thermal pollution problem is slight. This problem holds more importance for plant efficiency in relation to the possibility of drawing warmer water into the intake. Harleman[186] discusses the considerations of importance in design and analysis of such a situation. However, unless recirculation becomes significant and causes a continuous rise of cooling water temperatures, effects on the extent of heated water are minor.

Wedges Upstream from Diffusers

Diffuser systems have the purpose of thoroughly mixing the discharge water and hence eliminating any downstream surface layer. However, an upstream wedge may be formed if the mixed temperature is above the ambient temperature and the Froude number is sufficiently low. Establishing the local Froude number in the lower layer to be critical over the diffuser, the depth h_{2d} is found as[187]

$$h_{2d} = \left[\frac{Q_R - Q_o}{L_d \sqrt{g\left(\frac{\Delta p}{p}\right)_{Mo}}} \right]^{2/3} \quad (52)$$

where

Q_R = total flow in river

Q_o = diffuser flow
L_d = length of the diffuser
$(\Delta\rho/\rho)_{Mo}$ = density ratio based on temperature difference $T_M - T_R$

T_M = fully mixed stream temperature
T_R = ambient stream temperature

The condition for no upstream wedge is obtained by setting h_{2d} equal to the total available depth and finding the required river flow. The general solution for determining a wedge length can proceed in a manner similar to Bata's work discussed earlier.

In an outfall designed for complete mixing, care should be taken to evaluate the possible extent of any upstream wedge. However, it must be noted that the temperatures in the wedge are near the mixed temperature, T_M, not the outlet temperature, T_o.

Other Influences on Wedges
Cooling Effects

The two layer systems treated thus far have not been subject to heat exchange with the atmosphere. If the wedges are of sufficient length, however, loss of heat from the upper layer to the atmosphere may be significant. Since this cooling increases the density of the upper layer, (and hence its Froude number), the length of the wedge is decreased, and the interface position differs from that predicted by a conservative model.

A heat balance requires that the rate of heat transport into the wedge at the discharge point be equal to the rate of heat loss through the total surface area of the wedge. On a differential basis, the decrease in the rate of heat transport at a point must balance the rate of surface heat loss at that point. Harleman and Stolzenbach[188] have suggested the following formulation

$$q_t = K_E (T_1 - T_o) \quad (53)$$

where

q_t = heat flux out of water body
T_1 = average temperature of upper layer
T_o = air temperature above the flow
K_E = evaporative heat loss coefficient

Writing the one-dimensional equation for con-

52

servation of energy yields

$$U_1 \frac{dT_1}{dx} = -\frac{K_E}{\rho C_p h_1} (T_1 - T_o) \quad (54)$$

where ρC_p is the specific heat per volume of water. Using continuity, and assuming $(K_E/\rho C_p)$ constant, channel width B constant, and $T_1 = T_1$ at $x = 0$, the following is obtained by integrating Equation 54

$$\frac{T_{10} - T(x_1)}{T_{10} - T_o} = 1 - \exp\left[\frac{BX_1}{Q_1} \cdot \frac{K_E}{\rho C_p}\right] \quad (55)$$

where $T(x_1)$ = temperature of upper layer at $x = x_1$ and Q_1 is the flow in the upper layer. Knowing the temperature at succeeding sections of the wedge enables solution for wedge shape by numerical means.

Geometry Effects

The preceding theoretical work has treated a rectangular, prismatic channel. Natural geometric constraints may modify these results. Chikwendu and Francis[189] experimentally treated the problem of a curved channel, locating their outlet at the midpoint of a circular arc. Operating with densimetric Reynolds numbers Re_Δ between 10^3 and 10^4, they found the wedge length (centerline) to be a function of Re_Δ, but indicated the belief that this dependence would vanish for the larger values expected in natural rivers. Their data indicated the following expressions for centerline wedge length

$$\frac{L}{H} = C\,Re_\Delta^{0.9}\,F_\Delta^{-4} \quad (56)$$

F_Δ = densimetric Froude number
C = constant $(1.05 - 1.10)$ x 10^{-4} for curved channel runs = $(1.95 - 2.05)$ x 10^{-4} for straight channel runs.

It can be seen that the wedge length in the curved channel is only about half that in the straight channel. The helical circulations generated and observed in the curved channel were credited with causing this reduction. The mixing induced may create a less dense surface layer, thereby shortening the wedge. The wedge was also found to be longer on the inner channel wall than on the outer, implying an oblique frontal edge.

Mixing of Heated Waters by Momentum Jets

Jet diffusion has been studied by numerous investigators. Only those papers that are concerned with the buoyant (heated) jet will be discussed in this section.

Numerous factors play a role in determining the processes of jet actions. Chief among these are buoyant forces (created by density differences between the jet and receiving fluids), stratification within the receiving fluid itself, currents within the receiving fluid, and the finite physical bounds of the receiving fluid. The discharge of a jet into a denser receiving fluid will result in the jet being deflected upward due to the resulting buoyant force. This buoyant force is much less important than the inertia or momentum flux in the immediate vicinity of the jet efflux, or the zone of flow establishment. Then the trajectory which the jet describes becomes a function of both the initial momentum flux and buoyancy flux as the flow is decelerated.

The flow for a case with both initial buoyancy and momentum is called a buoyant jet or a forced plume. Two limiting cases exist. For pure momentum flux with no buoyancy, a simple jet, or free jet results. For pure buoyancy flux, the flow is called a simple plume.

The buoyant forces can arise due to any density difference, whether the cause be temperature, salinity, dissolved matter, or whatever. It would also be observed that the problem of a lighter jet rising in a heavier fluid is precisely the same as that of a heavier jet falling in a lighter receiving fluid. The only difference lies in the nature of the limiting boundary: a free surface in the former case and a solid boundary in the latter. Fan and Brooks[190] note that they found it more convenient to inject a heavier fluid into the lighter, noting that it makes no difference which way the apparent gravity acts since similarity is according to Froude's law. They also conclude effects at the free surface are negligible; it can be simulated by the solid bottom of a tank. This is likely true for deeply submerged jets, though added work, such as that by Maxwell and Pazwash,[191] may reveal more significant effects on shallow or surface jets.

Morton,[192] Abraham,[193, 194] Cederwall,[195] Frankel and Cummings,[196] and Fan and Brooks[197] have studied the problem of jets with both initial momentum and buoyancy. Fan[198] and Wiegel[199] review the results of several of these works.

If a jet rises due to buoyant action, it may extend to the free surface and spread out there. Frankel and Cummings[200] have suggested the addition of two more zones of flow to the zones of flow establishment and established flow.[200] These are the surface transition zone and the zone of horizontal surface flow. In the former, mixing occurs but with little dilution (change of density); in the latter zone, the flow spreads out radially and is subject to diffusion and mixing by waves, currents, and the like.

If the receiving fluid is not homogeneous, but is stratified, then the jet may or may not reach the water surface. Rawn, Bowerman, and Brooks,[201] Hart,[202] Fan,[203] Abraham,[204] Morton, Taylor, and Turner,[205] and Morton[206] have all performed work on jet discharge into stratified fluids. Wiegel[207] describes the three cases which can occur. The jet may reach the surface and there spread out. In a stratified receiving body, Fan[208] notes that as the rising plume becomes denser and the ambient fluid less dense, the buoyancy force acting on the plume will eventually reverse itself. The plume will then stop rising when the upward momentum vanishes due to the negative buoyancy force. This process leads to the other two cases Wiegel discusses: a jet which halts beneath the surface and spreads out, or a jet which has sufficient momentum to reach the surface and then plunges beneath the surface, where it spreads out as a submerged field.

All the above mentioned phenomena are modified by the existence of currents in the receiving fluid. Buoyant jets have been considered by Csanady,[209] Bosanquet, Horn, and Thring,[210] and Priestly,[211] as well as by Fan,[212] who also discusses the work of Callaghan and Ruggeri[213] and Vizel and Mostinskii.[214] Zeller[215] has made field studies in such a case.

Dimensional Considerations

Dimensional analysis has revealed important parameters for the jet problem, formalizing some of the factors which have been discussed. Rawn et al.[216] among others, gave the following general expressions

$$\frac{u}{u_o} = f_1 \left(\frac{S}{d_o}, \frac{r}{d_o}, \frac{\Delta \rho}{\rho_o}, F, Re \right) \quad (57)$$

and

$$\frac{C}{C_o} = f_2 \left(\frac{S}{d_o}, \frac{r}{d_o}, \frac{\Delta \rho}{\rho_o}, F, Re \right) \quad (58)$$

where

S = distance from the jet outlet measured along the jet axis
r = distance normal to jet axis
d_o = outlet diameter (or characteristic dimension)
u_o = axial jet velocity at $s=0$
ρ_o = density of jet fluid at $s=0$
ρ_s = density of receiving fluid
$\Delta \rho = \rho_s - \rho_o$ (sign reversed if effluent more dense)
F = densimetric Froude number of jet
Re = Reynolds number of jet
C = concentration in jet
C_o = concentration at $s=0$.

The concentration term is a measure of dilution of the initial jet, whether this implies dilution of a sewage effluent, decrease of temperature due to mixing, or the like. Several terms and means of expression for this dilution have been used. Abraham[217,218] defined

$$C = \frac{p - \rho_s}{\rho_o - \rho_s} \quad (59)$$

where p is the density at the given point under consideration. From this, noting that for $C = C_o$, $C_o = 1$, comes

$$\frac{C}{C_o} = \frac{p - \rho_s}{\rho_o - \rho_s}. \quad (60)$$

Some early work on jets from submerged marine sewage outfalls replaced the density terms in Equation 60 by concentrations to receive an equivalent expression for C/C_o. Frankel and Cummings[219] used such a formulation. The inverse of this was used by Rawn et al.[220] for data by Rawn and Palmer[221] and is called the dilution, S. All these terms simply relate the conditions in the jet to initial conditions and must, therefore, be related to the entrainment of fluid discussed earlier and thus to the flow dynamics.

Observing the density terms in the concentration expressions, it is convenient to write an expression involving temperatures, which are the ultimate concern in the heated discharge problem. Jen, Wiegel, and Mobarek[222] have done this as shown

$$\frac{T - T_w}{T_o - T_w} = f_3 \left(\frac{x}{d_o}, \frac{y}{d_o}, \frac{z}{d_o}, \frac{\Delta\rho}{\rho_o}, F, Re\right) \tag{61}$$

here

T = temperature at specified point
T_w = temperature of receiving fluid
T_o = temperature of jet at s=0
x, y, z = Cartesian spatial coordinates of point where T is to be evaluated.

Burdick and Krenkel,[223] in reviewing studies on buoyant jets, note little apparent effect for R_e as low as 2,000 to 4,000. It seems, therefore, that the Reynolds number can be neglected as a significant parameter in the jet problem so long as the jet may be considered turbulent.

Mathematical Treatment

The most frequently used approach to the treatment of jets subject to buoyant forces was introduced in 1955 by Morton et al.[224] and will be referred to as the Morton integral approach.

1. The fluids are incompressible.
2. Flow is fully turbulent, implying no Reynolds number dependence and that molecular diffusion is negligible compared to turbulent diffusion.
3. Longitudinal diffusion is negligible compared to lateral diffusion.
4. The largest variation of fluid density throughout the flow field is small compared with the reference density. Therefore, the density variation need be considered insofar as it gives rise to buoyant forces and can be neglected in inertial terms. The assumption of small density variations is commonly called the Boussinesq assumption.
5. Velocity and buoyancy profiles are similar in consecutive transverse sections of the jet.
6. The rate of entrainment is proportional to the magnitude of the difference between some characteristic velocity in the jet and the velocity of the ambient fluid unaffected by the jet. Morton notes, that ". . . if the mutual entrainment is turbulent, the only quantity determining the motion

is the relative velocity of the two streams."

Having made all the above assumptions, Morton then applied the equations of
conservation of volume flux,
conservation of momentum, and
conservation of density deficiency.
The resulting three equations, assuming Gaussian distributions for the similar profiles, are

$$\frac{d}{d_s}(b^2 u) = 2\alpha bu \tag{62}$$

$$\frac{d}{d_s}(b^2 u^2) = 2b^2 g \frac{\rho_a - \rho}{\rho_1} \tag{63}$$

$$\frac{d}{d_s}\left(b^2 ug \frac{\rho_a - \rho}{\rho_1}\right) = 2b^2 u \frac{g^d \rho_a}{\rho_1 dx} \tag{64}$$

where

u = a local characteristic velocity
b = a local characteristic length
s = axial coordinate of jet
α = entrainment coefficient, function of the shear due to the difference between the jet and ambient velocities
g = gravitational constant
x = distance
ρ = density in jet
ρ_a = density in ambient fluid
ρ_1 = reference density

Note that the right hand term in Equation 64 only exists if the ambient fluid is stratified.

Specifications of boundary conditions enable solution of the three equations for the three unknowns: u, ρ, and b. The equations take no account of currents in the ambient fluid.

Experimental Methods

There are two primary modes of experimental study: field measurements and data from lab modeling. The latter can range from an "infinite" receiving fluid to the development of a physical model corresponding to the complex combination of boundary geometries—and flow fields at some specific site.

Laboratory Modeling

The phenomenon of a heated jet can be studied by injection of a lighter fluid into a denser receiving fluid or vice versa. The cause of the density difference need not be heat, for the physical flow process is a function of the buoyant forces produced, not the agent producing them. Therefore, numerous techniques are available and work on the study of sewage outfalls and resulting plumes is useful.

In general, it is assumed that the important parameters such as the Reynolds number, Froude number, and current ratio, if the same in the laboratory as in the field, produce results in the laboratory which relate to field conditions. This is also true for such parameters as submergence, studied by Hall and Pazwash.[225] However, it is physically impossible to maintain complete similitude of both the Froude and Reynolds numbers if water is used in both model and prototype. This difficulty arises because of the very large $\Delta\rho$ required in the model. Hence, laboratory studies tend to model with Froudian similitude and to operate at Reynolds numbers sufficient to insure turbulent flows, which then lessens the dependence on Reynold's numbers.

Modeling for a Specific Site

In some cases where present experimental evidence is inapplicable and theoretical developments inadequate, it may be necessary to model the geometry and flow conditions expected at a particular site. Ackers[226] and Silberman and Stefans[227] have reported work of this type. Their work and a more detailed description of modeling techniques will be covered in a later section.

Field Studies

Zeller[228] conducted field work on the jet problem and discussed several alternative measurement techniques.

Drogues: Gross jet definition. Drogues can be used to trace the gross streamline pattern in heated jets, enabling determination of direction and a gross measure of velocity. Location of the drogues during the course of a single run is accomplished by surveying equipment, with the time of each reading noted. It was necessary to place all desired drogues simultaneously, rather than completely tracing one group and then placing another set near the outfall and repeating. The streamlines from the series of such trials yielded inconsistencies due to the tendency for the jet to meander during the course of a survey. Therefore, use of drogues implies use of a large enough number (released simultane-ously) to adequately define the jet, while at the same time being few enough to assure proper tracking. The path of the drogues also serves as a guide to desired locations for other measurements of jet feature, such as velocities, temperatures, and dye concentrations.

Use of temperature profiles. Use of temperatures to define the extent of a heated jet has been employed successfully both in the laboratory and in the field. Zeller used an underwater thermometer for measurements in conjunction with his drogue studies. Although some of his data did not give good definition, this was attributed to data-collection techniques.

Use of velocity measurements. Current meters may be used to measure velocities within the jet, and, particularly in the initial regions, to define its characteristics and approximate extent. The extensive use of velocity profile measurements can become burdensome and expensive. More important is the fact that velocity fields may not adequately define the region of interest due to more rapid dissipation of velocity than temperature.

Dye studies. Dyes have been used in the laboratory as a visual guide to locate the jet for profile measurements. Zeller made a few attempts to use a fluorescent dye, with concentrations measured by a fluorometer. However, lake depths of less than one meter near the outfall caused much of the dye to become attached to some of the numerous dense patches of vegetation in that vicinity. Hence, results were poor.

Dyes can be injected continuously or as a slug. The former would allow more time for measurements if the flow conditions and hence the jet remained unchanged. Due to the unsteady nature of river currents and the like, the slug injection more nearly yields the instantaneous movements rather than some time average of them.

Tracers other than dyes can be used. Harre-moës,[229] studying design of diffusers for sewage disposal in the sea, reported on the use of radio-active tracers to study jet diffusion and dispersion of sewage fields. Studies involved both submerged sewage fields and surface spread of a buoyant sewage field.

Infra-red techniques. Ackers notes the high cost of extensive data collection by conventional means; difficulty exists in obtaining sufficient data over a short enough time span to fairly represent instanteous jet conditions. Use of airborne infra-red techniques has, therefore, been advocated for obtaining surface isotherms. Experience in the United Kingdom shows that scanning by an infra-red sensor in an airplane provides a close detail of surface temperatures, and methods exist for conversion of the data into isothermal maps.[230]

Zeller also discussed infra-red surveys, though only one series of six scans was used. The data collected were limited, although a general view of gross jet movement was provided. As instrumentation, techniques, and methods of data reduction improve, infra-red techniques may prove important for surface jet studies.

State of the Art

Zone of Flow Establishment in Quiescent Receiving Waters

The definition of the first stages of a jet requires treatment of the zone of flow establishment, where the potential core is considered to still exist. Mixing finally reaches the center, so that at the end of this zone the Gaussian (or whatever lateral profile is considered to hold) profile has developed. For discharge into an essentially quiescent fluid, the results of Albertson et al.[231] are frequently used. They show C_2 in Equation 65 as equal to 0.081, implying that the zone of establishment ends at 6.2 d_o from the jet discharge

$$\frac{X_o}{d_o} = \frac{1}{2C_2} \qquad (65)$$

where

$X_o =$ limit of zone establishment
$d_o =$ diameter of circular orifice
$C_2 =$ constant

Reynolds has prepared the tabulation of C_2 values shown in Table 8[232] (see page 45).

Fan[233] has suggested use of Albertson's value also for inclined buoyant jets, assuming that buoyant forces play a minor role in these early stages of the jet. Based on this assumption, the half-width, b_o, and the dimensionless concentration, C_o, at the end of flow establishment can be defined as below

$$b_o = \frac{d}{\sqrt{2}} \qquad (66)$$

$$C_o = \frac{1 + \lambda^2}{2\,\lambda^2} \qquad (67)$$

where λ, the turbulent spreading ratio, is considered by Fan to be 1.16, as found by Rouse, Yih, Humphreys, et al.[234] The solution for the jet in the zone of established flow can then be applied as beginning at these values of b_o and C_o.

Effect of Currents

Fan has examined results of other workers for non-buoyant jets in a cross stream and assumes it also valid for the buoyant case.

Fan then finds b_o by considering the conservation of the flux of density deficiency,

$$b_o = d_o \sqrt{\frac{k}{k + \text{Cos}\,\theta_o}} \qquad (68)$$

where $k =$ velocity ratio.
$\theta_o =$ angle of jet to the x axis at end of zone of establishment.

Pratt and Baines[235] have studied the zone of establishment and showed the strong influence of cross streams on the length of establishment and indicated that even a small stream current can reduce this length appreciably. Thus, for k equal to 3 or 4 (typical values for a discharge canal), S_e'/d_o is 3 or less where S_e is the length of the establishment zone along its axis. For discharges into streams of higher velocity, where k approaches 1 to 2, the zone length would approach the order of the discharge width. For discharge through pipes or a diffuser, k values should be greater, implying a greater length of the zone of establishment. This implies mixing with a greater volume of fluid, which is usually the goal of such diffuser systems.

Rates of Fluid Entrainment

Effect of Cross Streams

The apparent effect of cross streams is to increase the magnitude of the entrainment coefficient. Keffer and Baines[236] found it to increase along the jet axis, ranging from 0.3 initially to as high as 1.6 at 10 diameters downstream. The values decreased as k increased; this, of course, approaches the case of no ambient current, where α is about 0.082. Keffer and Baines did not use a vector velocity difference. Fan[237] did

TABLE 7

COMPARISON OF STREAM TEMPERATURE PREDICTION MODELS

Gameson, Gibbs, and Barrett

$$\frac{d\theta}{dt} = -\frac{f}{z}\,\theta,$$

where θ = excess of water temperature over natural water temperature, °C

f = exchange coefficient, cm/hr

z = mean river depth, cm

Velz and Gannon

$$\frac{dT_w}{dt} = -\frac{H}{62.4b}$$

where T_w = water temperature, °F

H = rate of heat loss from water surface, $\dfrac{BTU}{ft^2 \text{-} day}$

b = river depth, ft

Duttweiler

$$\frac{dT}{dt} = \frac{1}{\rho c}\,\frac{\lambda}{z}\,(T_E - T)$$

where T = water temperature, °F

λ = parameter dependent on atmospheric conditions,

z = hydraulic depth, ft

T_E = equilibrium temperature, °F

Edinger and Geyer

$$\rho\, C_p U d\,\frac{\delta T}{\delta x_1} = -K(T - E)$$

where T = water temperature, °F

U = mean stream velocity, ft/hr

d = mean stream depth, ft

K = exchange coefficient, $\dfrac{BTU}{ft^2\,day}$ °F

E = equilibrium temperature, °F

C_p = specific heat water, $\dfrac{BTU}{lb}$

ρ = density of water, $\dfrac{lbs}{ft^3}$

---

use such a difference, and found values of α between 0.4 and 0.5 for k values from 16 to 2, respectively. These were found by choosing a value which gave best fit to observed profiles; however, another coefficient (a drag coefficient for the pressure difference across the jet) was being simultaneously manipulated to give best fit. Therefore, Fan's values are most indicative when taken as an order of magnitude.

Zeller,[238] studying a jet discharged on the water surface, also found values of α ranging from 0.496 for k = 1.75 to 0.063 for k = 9.7,

with α decreasing relatively consistently as k increased. The values must be considered carefully, however, for Zeller's definition differs from virtually all others working with cross streams. Rather than considering a velocity difference, he uses only the jet centerline velocity.

Vertical Entrainment

The preceding work has assumed that fluid is equally entrained on all sides of the jet. However, the rate of entrainment through these surfaces may not be equivalent. One example is a free surface, where no entrainment occurs. At the lower surface of a surface jet, shear action tending to entrain fluid is being countered by gravity acting through density differences. It might be expected, therefore, that vertical entrainment would occur in the early reaches of the jet, where shear is sufficient to overcome buoyancy, and then decrease (as the shear decreased) until finally no vertical entrainment occurred. The work of Ellison and Turner,[239] which allowed only vertical entrainment, showed such results. For this case, they formulated α as a function of R_i, the Richardson number which is equal to the inverse of the square of the densimetric Froude number, finding $\alpha = 0$ for R_i greater than 0.8. This corresponds to a densimetric Froude number less than about 1.1, indicating that inertial shear has become small compared to buoyant forces. Zeller's data[240] seem to verify the reduced vertical mixing at later reaches of the jet; however, the initial reaches revealed nothing about such mixing as the jet occupied essentially the entire depth of water for the first several hundred feet from shore. Zeller assumed no vertical mixing (which implies a constant jet thickness) as giving the best fit for his model. He later noted that measurement inaccuracies could have masked gradual increases in thickness.

Jen et al.[241] and Wiegel et al.[242] have studied surface jets, and their data reveal a definite vertical entrainment in early stages of the jet and decreasing gradually. Jen found that the bottom of the mixing jet, where initial densimetric Froude numbers varied from 18 to 180, has a slope between 0.125 and 0.182. He noted that the surface jet mixed more laterally than with depth, varying by a factor of about two.

Longitudinal and Lateral Variability of α

The entrainment coefficient varies, both from side to side and in the axial direction, in lateral mixing. Most studies have neglected such effects and merely used an average, constant α. Sawyer[243] used an overall α, but allowed for differing rates of mixing, α_1 and α_2, on the inside and outside of his two-dimensional jets. Using a sech^2 velocity profile, Sawyer found $\alpha = 0.130$ satisfactory for both cases. He also found mixing occurred on the outer edge of the jet at twice the rate of the inside edge.

Keffer and Baines[244] made measurements of mass flux as it varied along the jet axis and, observing the gradient of mass flux, computed values for α of 0.2 to 1.8. As expected, α increased with an increase in k, the velocity ratio. Keffer and Baines dealt with a three-dimensional, round air jet discharged into air. They made no attempt to predict gross jet characteristics by use of an average α. It should be noted that the presence of buoyant forces inhibiting vertical entrainment could alter the relations shown by Keffer and Baines.

Zeller developed an analytical model showing α as a function of the axial distance, s, and the jet curvature and verified it with field data. It indicates an almost linear relation between α_v and axial distance. Zeller made analyses based on use of a constant α and also based on a varying α and found a negligible difference despite α_v/α as high as 1.52 some 1800 feet along the jet.

Jet Definition: Size and Path

Effects of Cross Streams

Bosanquet et al.[245] studied the paths of a jet discharging into stagnant ambient fluid, providing theoretical analyses which yielded reasonable agreement with observed paths. Keffer and Baines studied jets directed normal to a cross stream. Their results showed more rapid curvature for lower values of k, as expected. For (s_o, Z_o) representing the end of flow establishment, $(s - s_o)/dk^2$ versus $(s - s_o)dk^2$ yielded a single curve for $k > 4$. Use of k^2 was justified because in the absence of pressure fields, the jet path depends on the ratio of its vertical momentum flux to that of the cross stream, or k^2.

Fan[246] considered a round, submerged, buoyant jet, with Gaussian profiles for both velocity and density deficiency. He also noted that a higher pressure can be expected on the upstream edge of the jet than the downstream edge, thus exerting a drag force on the jet plume. Conservation of momentum in the x-direction then becomes a function of momentum flux due to entrained fluid and the drag force; while in the y-direction, the gravity force on the cross section minus the y-component of drag equals the rate of change of momentum flux. Fan also derived two equations for conservation of density deficiency and flux of a tracer of concentration C.

The equations can be reduced to a more tractable form using the assumed forms of the transverse profiles. If values for the entrainment and drag coefficients could be obtained, a solution for the jet characteristics would be possible. Fan obtained some values for these coefficients by the best fit or experimental trajectories and profiles.

Effects of Wind Stresses and Tides

Wind stresses may, depending on direction, increase or dissipate the jet's momentum flux. Assuming jet velocity negligible compared to wind velocity, Zeller expressed the wind stress by

$$T_W = \rho C_W W^2 \qquad (69)$$

where W is a characteristic wind speed and C_w is a drag coefficient.[247] For Lake Monona, wind stresses were found to be usually quite low. However, studies by Cornell University at Cayuga Lake[248] indicate that wind stresses there were of sufficient magnitude to have appreciable effect on movement of surface jets.

The effect of tides on power plants located on estuaries, bays, etc. is not well known because of the unsteady flows. Ackers reports[249] on several such problems in Great Britain investigated by modeling.

Effect of Previously Stratified Fluid

Fan[250] modified Morton's integral method to consider varying initial angles into stratified fluids using a constant α throughout the entire jet flow region. This solution is especially valuable for jets discharged with a strong horizontal momentum, as expected for most subsurface heated discharges. Using $\alpha = 0.082$, Fan found good agreement with measured jet trajectories and boundaries for these cases. He found three relevant parameters for the problem: the initial jet angle, the jet densimetric Froude number, and a stratification parameter.

Effects of Outlet Characteristics

The initial jet characteristics, width, depth, and velocity, will have considerable influence on the absolute scale of jet path and growth. These effects can be expressed consistently by normalizing the variables by dividing by the initial values. However, other factors, such as shape of the outlet and its orientation, need study.

Wiegel[251] indicated that rectangular surface jets mix more in the manner of a jet from a circular nozzle than from a two-dimensional slot. The efficiency varies, however, with the exact shape.

Orientation of the jet outlet has been treated analytically by Fan[252] for round, submerged jets. Widths, depths, and paths for various combinations of initial and receiving conditions can be investigated through his results. Studies of sewage outfalls provide some insight into possible use of subsurface discharge of heated waters.

Dispersion of Heated Discharges

Much work has been done on dispersion and diffusion of substances of the same density as the receiving water. Work has been done on these processes when density differences do exist.

The basic dispersion-causing mechanism is the difference in velocity in different parts of the flow cross section. The part of the dispersing material in the regions of higher velocity is carried ahead of the portions in slower regions. This spreading tendency is somewhat opposed by turbulent diffusion, laterally and vertically, from the faster- to slower-moving regions. Such diffusion also contributes slightly to the longitudinal spreading. The basic equation for conservation of mass, assuming only velocity in the x-direction, is

$$\frac{\partial c}{\partial t} + u\frac{\partial c}{\partial x} = \frac{\partial}{\partial y}\left(e_y\frac{\partial c}{\partial y}\right) + \frac{\partial}{\partial z}\left(e_z\frac{\partial c}{\partial z}\right) + \frac{\partial}{\partial x}\left(e_x\frac{\partial c}{\partial x}\right) \qquad (70)$$

where

c	= concentration
x, y, z	= longitudinal, vertical, lateral coordinates
u	= longitudinal velocity
e_x, e_y, e_z	= turbulent mixing coefficients

The last term in Equation 70 is frequently neglected.

Taylor[253] proposed a simplification expressing the dispersion process by a one-dimensional Fickian diffusion equation as

$$\frac{\partial C}{\partial t} = D_L\frac{\partial^2 C}{\partial x^2} - \bar{u}\frac{\partial C}{\partial x} \qquad (71)$$

where

C = cross-sectional mean concentration
$\bar{u}$ = cross-sectional mean velocity
D_L = longitudinal dispersion coefficient

Considering only flow in a pipe, Taylor found the coefficient in this equation to be

$$D_L = 10.1 \, a \, U_*$$

where

a = pipe radius
U_* = friction velocity = $\sqrt{\rho_o/\rho}$ (72)
ρ_o = wall shear

Elder[254] found that for flow in an infinitely wide, two-dimensional open channel

$$D_L = 5.9 \, d \, U_* \qquad (73)$$

where d = flow depth.

As noted by Fischer, however, dispersion coefficients observed in natural streams have been much larger than the range shown by equations 72 and 73 varying from 50 to 700R U_* where R = hydraulic radius.[255] Fischer[256] noted that the primary cause for this trend seemed to be lateral non-conformity, especially lateral velocity gradients. His analysis of the flow by extending Aris' method of moments demonstrated that lateral gradients increased D_L several orders of magnitude. The precise form of the velocity field thereby becomes important. Fischer[257,258] has shown that Equation 71 is valid only after an initial period during which convection dominates the dispersion. Equation 71 is valid only when the variation of concentration across the section is small. The distance, L, from the point of tracer injection, required for validity of equation 68 is

$$L > 1.8 \frac{l^2}{R} \frac{\bar{u}}{U_*} \qquad (74)$$

where l = distance from point of maximum surface velocity to farthest bank. Variations in coefficient values computed by different methods prior to reaching L are noted by Fischer.

Determination of D_L

The most frequent means of finding D_L had been based on tracer studies at the site of interest. A solution to Equation 71 is given by Thackston and Krenkel[259] and many others

$$C = \frac{M}{A \sqrt{4\pi D_L t}} \exp - \frac{(x - \bar{u}t)^2}{4 D_L t} \qquad (75)$$

where

A = cross-sectional area of flow,
M = weight of injected tracer material

A number of solutions to Equation 71 are given by Parker.[260]

Fischer[261,262] describes a method of obtaining dispersion characteristics for a given stream without the need for tracer studies, which are costly.

$$D_L = - \frac{1}{A} \int_o^b q'(z) \, dz \int_o^z \frac{1}{e_z \, d(z)} \, dz \int_o^z q'(z) dz \qquad (76)$$

where

b = channel width
d(z) = depth at lateral point z

$$q'(z) = \int_o^{d(z)} u'(y, z) \, dy$$

u'(y,z) = deviation from cross-sectional mean velocity.

The lateral turbulent mixing coefficient, e_z, is evaluated from Elder's work:

$$e_z = 0.23 \, d \, U_* \qquad (77)$$

Evaluation of D_L then requires field measurements of stream slope and the geometry and velocity distribution in one or more typical cross sections. Fischer notes that a crew of three men can usually obtain such measurements in less than a day. Values calculated by this method have been compared with previously published work.[263] Even in non-uniform streams, D_L was accurate within a factor of 4; thus, the length of the dispersing cloud and peak concentration decay rates can be predicted within a factor of 2. In uniform streams D_L was usually predicted within 30%.

In a later paper, Fischer has considered the effect of bends and meanders on longitudinal dispersion and transverse mixing. The transverse coefficient, e_z, given by Equation 76, has been shown adequate for uniform flows in field conditions by Fischer. Values for curving flows can be somewhat higher due to secondary currents induced in a bend. Adapting an earlier analysis, Fischer[264] develops an expression for e_z, showing e_z proportional to $\overline{u}^2d^3/R_b^2U_*$, where R_b is the radius of curvature. Its use is limited by assumptions in its derivation and uncertainty in the needed parameters. It is suggested, however, that one find, for the given stream, the maximum e_z which can be obtained by the equation. If this is less than 0.23 d U_*, then secondary flow is probably not significant, and the uniform flow results are adequate. If e_z is found higher than 0.23 d U_*, then the coefficient must be evaluated experimentally.

Application to Heated Discharges

Edinger and Polk[265] have provided solutions for the case of a heated discharge into a river of uniform velocity. The discharge is located along the bank and the efflux is parallel to the stream flow. The assumption is made that the jet does not affect the main stream flow and Equation 70 is reduced to the following steady case

$$u\frac{\partial c}{\partial x} = \frac{\partial}{\partial z}\left(e_z\frac{\partial c}{\partial z}\right) \qquad (78)$$

Two other solutions are presented: the case above with surface cooling considered, and the case in which vertical (y-direction) mixing is also important. By plotting field data versus the curves, some insight can be gained into the important mechanisms for the given case.

Two means of describing the data have been used: the fraction of cross-sectional area contained within a given temperature rise contour (as a function of distance from the source), and the surface area contained within a given temperature rise contour. The reader is referred to

the report for details. Some of the general analytical findings follow.

1. The surface area varies inversely with the cube of the concentration ratio for the two-dimensional case and the 3/2 power for the three-dimensional case.

2. For a given concentration ratio, the area varies with the cube of the plant pumping rate, providing a means for evaluating the effects of plant growth. This also becomes the 3/2 power for the three-dimensional case.

3. For a given concentration ratio, the area varies inversely with the square of the river flow in the two-dimensional case.

4. (2-dimensional)

$$L_M = \frac{bQ_R}{\pi e_z d}\left[1 - \left(\frac{Q_P}{Q_R}\right)^2\right] \qquad (79)$$

where

L_M = distance to fully mixed contour
Q_P = plant flow
Q_R = river flow

(3-dimensional)

$$L_M = \frac{Q_R}{\pi\sqrt{e_z e_y}}\left(1 - Q_P/Q_R\right) \qquad (80)$$

It is necessary to test the theory more adequately with field data to determine its usefulness. One question to be answered is how much initial momentum limits the value of the results.

Of course, similar solutions can be developed for other outlet arrangements and orientation, but the turbulent mixing coefficients e_y and e_z need evaluation. These have not been defined for heated flows.

MODELING OF HEATED DISCHARGES

The basic requirements of modeling for homogeneous fluids have been described in numerous texts and articles. The reader is referred to Keulegan[266] for a description of the dynamic, dimensional, and equational methods of deriving model laws for use for distorted models of saline wedges.

Stages of Stratified Flows

Ackers[267] discusses some general considerations in modeling of heated effluents, noting that a first requirement is similarity of flow pattern in the receiving fluid in the case where motion already exists prior to jet discharge. This modeling is ordinarily obtained by equating Froude numbers (not densimetric) in model and prototype. The main stages of dispersion of the heated discharges are:

a. Turbulent entrainment at the efflux jet. Close to the outfall, the inertia of the jet is important while density differences are not.

b. Buoyant rise of jet, if submerged. The trajectory of the jet is dependent on initial inertia as well as buoyancy force, the mixing being due to entrainment by turbulence at the plume boundary.

c. The convective spread of effluent from such an initial dilution zone over the surface of the receiving water. This process is dependent on the density difference where the convective spread has its origin, e.g., from the boil point of a submerged jet. Whether or not mixing will occur at the interface is dependent on the densimetric Froude number.

d. The mass transport of the effluent by ambient currents, which impose additional velocity vectors on all other stages; these will vary periodically in tidal areas.

e. Diffusion and dispersion due to the turbulence in the ambient fluid.

f. Loss of heat by evaporative cooling.

It is perhaps best to discuss modeling principles as related to these stages of the process. Both stages a and b must be operated at a natural scale (horizontal scale equals vertical scale). Vertical distortion of the model would cause loss of similarity with regard to the vertical turbulent mixing occurring.

For studies of convective spread, stage c, Barr[268] has developed the densimetric Froude-Reynolds number for a geometrically similar model to specify equality in model and prototype. Keulegan[269] called this term, the densimetric Reynolds number.

$$\overline{F\Delta R} = \frac{(\frac{\Delta \rho}{\rho}\ g)^{1/2}\ l^{3/2}}{\upsilon} \qquad (81)$$

where l = characteristic length.

Barr notes, however, that if turbulent flow is assured in both model and prototype that strict equality of this parameter is not needed. In fact, such models are usually established on the basis of the densimetric Froude law, with a densimetric Reynolds number high enough to provide turbulent conditions.

Other studies[270] have been made on convective spread, including modeling the rate of advance of the heated water front. Hooper and Neale[271] reported on an early study in the United States. Extensive modeling has been utilized in Great Britain, where tidal variations of flow provide added difficulties.[272-274] The extensive work on saline wedges by Keulegan[275] at the National Bureau of Standards has provided additional information. A recent U.S. model study with tidal action involved has been conducted by Harleman and Stolzenbach.[276]

The Froude law is used for stage d, where the ambient currents are superimposed on the jet discharge. The ambient currents can be treated by the standard laws for homogeneous fluids, i.e., the Froude number, assurance of a Reynolds number sufficient to insure turbulent flow, and scaling roughness to give correct head losses. Barr[277] reports experiments showing that in those cases where internal movements (density generated) outweigh gravity effects the Froude criterion can be lapsed, resulting in possible reductions in horizontal scale or vertical exaggeration (distortion).

Little has been done on stage e, diffusion due to ambient turbulence. However, a natural (undistorted) model must be used to maintain similarity of horizontal and vertical diffusion.

Frazer et al.[278] have shown that longitudinal dispersion will be simulated adequately if the dissipative energy is properly treated by adjusting model roughness. The case in point involves almost purely longitudinal gradients, and hence the distorted model was adequate.

The loss of heat by cooling (stage f) could have significant effects on wedge lengths and other features. Hence, similarity may be sought, though practicality often prevents strict similarity. If test durations are sufficiently short, for example, effects of cooling may be slight. However, some check of the possible effects should be made in any event. Harleman and Stolzenbach[279] observe that it would **not, in gener**al, be possible to have both similarity of frictional effects and heat loss. **Maggiolo and Spitalnik**[280] have considered models for **change of cooling** water temperatures with time, **as affected by** solar radiation. They relate **scales for relative** humidity, solar radiation, **and stored heat** to distorted models.

There is at least some background available for modeling all the stages of **heated water dis**charges. Some of the processes **can be modele**d simultaneously, while others **cannot. Modeling** of stages a and b can most **conveniently be** accomplished by using the **same initial density** difference in model and prototype. **Stage d is** also compatible with this **model. Ackers** also discusses the exaggeration of **the vertical** scale in the model to properly study **stage c and** the use of two models to enable **study of cases wh**ere stages a, b, c, and d are all of **importance. The** second model is a large-scale **one of the immedi**ate locale of the jet outlet, **due to difficulti**es of representing the initial jet mixing **in vertically** exaggerated models.

Distortion in Heated Discharge Models

Distortion in models of **homogeneous fluids** has been commonplace. Its **need arises due to** the effect of viscosity, which **places a limit on** the vertical scale. Keulegan[281] **has shown analyti**cally a means for finding **approximately this** limit. Essentially, if adherence **to this minimum** scale for horizontal dimensions **yields too large** a model, then distortion is **employed. Howeve**r, distortion, as mentioned earlier, **rules out simi**larity of vertical and lateral **turbulent mixing**. Where convective spread is **dominant, distortion** may in fact be necessary to **obtain similarity**. Ackers,[282] for a model using **the same fluid** and density difference as the prototype, **showed that** the minimum distortion required **for convec**tive spread similitude is the **following**

$$D^{5/2}\lambda_l{}^{3.2} \geq \cfrac{150}{\left[\cfrac{\sqrt{g\,(\Delta p/p)}\ d^{5/2}}{L\upsilon}\right]_p} \qquad (82)$$

where

L   = distance over which **spreading** occurs

λ_l = length scale
D   = ratio of depth scale to length scale
d   = depth of fluid
$g\,\Delta p/p$ = densimetric gravitational force
υ = kinematic viscosity

However, lesser distortions are acceptable if the error involved in convective spread to some critical position is small. This can be checked by use of congruency diagrams as recommended by Barr.[283] Barr has suggested considering first the limits of spread of the heated discharge. If the spread is not adequately simulated then other elements will not display similarity. He showed that, for the same density differences in the model and the prototype a small natural model could not yield similar limits of spread. The best degree of distortion is determined by using congruency diagrams for simpler modes of flow involving density spread. Choice of distortion is, thus, removed from the arbitrary category, though more work is needed to establish the limits of the proposed method. Examples of application of the congruency diagrams to change are given by Barr[284] and Ackers.[285] Fairly complete congruency diagrams are given by Frazer et al.[286] for both underflow and overflow conditions.

Physical Requirements of Models

Keulegan[287] demonstrates that if the Froude law establishes river discharges in a distorted model of saline wedges, the density scale must be unity. If the density difference in this model is chosen greater than the density difference in the prototype, Barr[288] states there must be a corresponding change of the surface slope (natural or distorted) directly proportional to the increase of the density difference in the model to simulate mixing of stratified flows. This is, then, a variance from the Froude law.

If at all possible, a unity density scale should be employed. If this is not possible with heated water, it might be desirable to try operating with a model using salt water as the receiving fluid. This may especially be true if very high temperature differences exist in the prototype or test durations are long, both resulting in high heat loss. For lower temperature rises (and lower density deficiencies) salt is more difficult to employ due to measurement limitations. The final choice of media must rely on considerations of the need to simulate heat losses.

Ackers[289] has noted the large heat requirement of large-scale models. For example, twelve hours of heating with immersion heaters permits only one hour of testing. Hence, the model program must be carefully planned, with automatic data collection very useful. Ackers[290] describes the control devices and instrumentation used at the Wallingford laboratories in Great Britain. The possibility of using infra-red scanning techniques to provide more rapid collection of surface isotherm data is also suggested.

COOLING PONDS AND RUN OF THE RIVER COOLING

Cooling Ponds

Cooling ponds are being increasingly used as a source of circulating water for central steam electric stations when land is available at a reasonable price, because they frequently are the simplest, cheapest, and least water intensive method of water cooling.[291] Ponds may be constructed by simply pushing up an earth dike, may be operated for extended periods without makeup water, and may serve as a retention basin. Ponds, unfortunately, also have low heat transfer rates, thereby requiring large surface areas. They concentrate dissolved solids, thereby requiring blowdown, and collect impurities since large surfaces are open to the atmosphere.

Possibly the most comprehensive discussion of heat transfer, movement, and ambient conditions in natural lakes and ponds is the chapter in Hutchinson[292] on the thermal properties of lakes. The natural thermal cycles of bodies of water are described and standard definitions presented. The thermal properties of lakes vary with latitude, altitude, and depth of the basin. In the temperate zone most lakes can be described as dimictic; i.e., are freely circulating twice a year, in spring and autumn, and inversely stratified in winter and directly stratified in summer. Lakes over a wide range of latitudes absorb during their complete annual cycle 110,000 to 150,000 BTU/sq ft/yr. Of this total, 60 to 70% is taken up after the spring overturn and would, of course, require work to be carried down into the hypolimnion. The absorbed summer dose is about one half the total radiative energy reaching the water surface.

Though there are many individual articles[293-299] describing cooling ponds, perhaps the most comprehensive treatment is in Berman's *Evaporative Cooling of Circulating Water*.[300] Though based on Russian experience, it appears to be the best collection of data about and analysis of cooling ponds available.

Berman suggested that cooling ponds be used only for heat loads greater than 800,000 BTU/hr. For smaller heat loads, tanks with surface areas of 8500 to 11,000 sq ft should be used. Suggested dimensions adopted from regulations of the Heating and Power Board of the Soviet Union for cooling ponds are:

Depth greater than 5.0 feet at low level

Depth greater than 8.2 feet at normal water level

Depth less than 10.0-13.0 feet since circulation ordinarily does not extend deeper.

For very large ponds, 1200-1700 acres, maximum depths are from 50-60 feet. In order to efficiently utilize the total area of the pond, stream guides may be provided.

According to McKelvey and Brooke,[301] earth dikes for cooling ponds can be only 6 to 8 feet high, constructed at a low cost, operate for extended periods of time with no makeup, and serve as a holding basin for contaminant detection.

The major disadvantages are the large surface areas required, 1 to 2 acres per megawatt of installed capacity; the possibility of fog, on cold days at distances up to 600 feet downwind from the pond; and the low heat transfer rate across the water surface. Brooke and McKelvey[302] indicate transfer rates of 3½ BTU/hr/sq ft/°F surface water and air. Berman[303] suggests a value of 75-150 BTU/hr/ft². Berman's value seems to be closer to the values generally found in the United States. Langhaar[304] has prepared a series of nomograms based on an energy budget which makes it possible to compute rapidly the surface required for cooling ponds.

In computing the surface area required, daily atmospheric conditions may be used if the holdup time in the pond is greater than 24 hours. The major factors affecting pond performance are surface air temperature, relative humidity, wind speed, wind fetch, solar radiation, silting of the pond, aquatic growth, and erosion. It should be recognized that the determination of the size of the pond is still an art because of the many anomalies.[305] Velz[306] has shown how the variability of atmospheric parameters may be taken into account.

Edinger and Geyer[307] have analyzed cooling ponds on the basis of their circulation patterns and temperature distribution. Though a number of ponds have been designed by rough rule of thumb, 1 to 2 acres per megawatt of installed capacity, a more sophisticated design would take into account the cold water intake rate, the amount of waste heat, and atmospheric conditions.

An approach* of 3 to 4°F is the lowest practicable amount of cooling in ponds of reasonable size. However, deep cooling ponds can provide water at colder temperatures during the year than can even draft towers in colder climates. Based largely upon the pioneering work of Harbeck,[308] it can be shown that the amount of water lost in cooling ponds to dissipate heat may be less than the amount lost in cooling towers. Though evaporative losses are usually the major losses, radiation and conduction can for the conditions listed dissipate up to 65% of the heat load.[309] Since wet cooling towers depend almost exclusively upon evaporation for cooling, it can be seen that the water losses in cooling towers can exceed water losses in cooling ponds. If, however, the ponds utilized are not previously existing ponds, then the normal evaporative losses must be added to the heat induced losses.

Berman[310] has suggested costs of $2 to $3 or more per cfs of cold water flow. Maine Yankee costs for cooling ponds[311] were estimated to be $300 to $500/cfs of cold water flow. These costs indicate the wide variation in costs for cooling ponds.

Spray systems added to cooling ponds increase the heat loss per unit area but also increase the water losses due to the enhanced evaporation. Spray ponds have many of the characteristics of cooling ponds; however, the heat transfer rate is higher, from 2600 to 5500 BTU/hr/sq ft,[312] according to the Russian experience, or 4000 to 10,000 BTU/hr/sq ft according to American and English experience.[313] Spray ponds, however, are usually unsuitable for a cooling range greater than 18°F. Spray ponds are also, in general, considered uneconomical for large size plants; as the size of the pond increases, the air passing by the sprays is already saturated so the sprays on the downwind side are not very effective. However, South Carolina Electric and Gas Company[314] has built a 10-acre spray pond in conjunction with a swamp cooling area to dissipate 13°F of heat from 180,000 gallons per minute. The capital cost was $985 per million BTU removed.

Run of the River Cooling

Until the present concern about thermal pollution became manifest, the availability of large supplies of cooling water was a prime requirement for large power plant locations. Most plants in the East, Midwest, and South used run of the river or pond cooling with the overwhelming proportion utilizing run of the river cooling. In 1965, in the North Atlantic Region, (as defined by FWPCA) only 1 plant (100MW) out of 101 utilized cooling towers, in the Southeast 2 out of 61, in the Great Lakes region zero out of 54, and, for the country as a whole, 116 out of 514 plants.[315] The plants utilizing towers were previously the smaller plants, less than 400 MW, and primarily in the West.

The situation has drastically changed since then and we now find units of 600 to 1000 MW commonly being ordered and station sizes of 3000 to 4000 MW. For an average condenser rise of 15°F, a fossil fueled plant of 4000 MW would require about 4600 cfs while a nuclear fueled plant would require about 7000 cfs. To limit the temperature rise to 5°F, the standard recommended by The National Technical Advisory Committee on Water Quality Criteria[316] would require 14,000 cfs and 21,000 cfs, respectively. The average annual flow of only a few rivers in this country exceeds this value.

Detailed comparison of the cost of run of the river plants with other cooling systems is given in the chapter on Comparison of Cooling Methods.

---

*Approach is the difference in temperature of the cool water leaving the pond and the ambient wet bulb temperature.

66

COOLING TOWERS

Major Sources of Information

Though a wide variety and large numbers of articles have been published on devices for cooling condenser water, the two best and most detailed technical references in the English language are *The Industrial Cooling Tower* by McKelvey and Brooke[317] and *Evaporative Cooling of Circulating Water* by Berman,[318] which is a 1961 translation by Raymond Hardbottle of the Russian book published in 1956. Though quite different in its outlook, one should also note the excellent study *Water Demand for Steam Electric Generation* by Cootner and Löf.[319] For a basic understanding of the integrated concept of electric power generation at an individual station comprising boilers, generators, condensers, and cooling towers and their effects on efficiency and costs it is unsurpassed. What the book does not do is to compare alternate systems at alternate sites. While run of the river cooling water may, in general, be cheaper than cooling towers, if the run of the river plant is not at the load center or at the fuel source, the integrated cost for delivered power may be less for a plant with a cooling tower at the load center or mine than run of the river plant at some more distant location.

For a basic understanding of the workings of a central electric power generating station, see *Power Station Engineering and Economy* by Skrotski and Vopat.[320] For a more detailed examination of the generation and production of steam from fossil fuels, the in-house publication of the Babcock and Wilcox Company — *Steam — Its Generation and Use*[321] is recommended.

Also useful as reference works are the house publications of the major manufacturers of cooling equipment and the Cooling Tower Institute. Representative publications include: *Cooling Tower Fundamentals and Application Principles*, Marley Company,[322] and *Cooling Tower Performance Curves*, Cooling Tower Institute.[323] Up-to-date articles on individual topics in the cooling water field are ubiquitous, and only the articles deemed to be most important are included herein.

A good pictorial introduction to the field, outlining basic components and concepts, is the special report, *Cooling Towers,* by Elonka.[324] A schematic drawing of the different types of cooling systems is given in Figures 12 and 13 (see pages 68, 69).

Definitions

Certain specialized definitions are required to read the literature in this field. They are

Cooling range or range: The number of degrees water is cooled from inlet to outlet of device.

Approach: The difference in temperature of the cold water leaving and the wet bulb temperature of the air entering. Without special treatment water cannot be cooled in a water cooled device to a temperature lower than the wet bulb ambient temperature. It has been suggested by Agnon and Young[325] that it is possible to reduce the wet bulb temperature of the incoming air below ambient by passing the incoming air through an air cooler or heat exchanger before being admitted to the tower proper. The coolant used was part of the chilled water from the tower.

Heat Load: Amount of heat dissipated per unit time in the cooling device. It is equal to the weight of water circulated per unit time times the approach.

The heat load can be increased by:

1. Increasing the velocity of the air in contact with the water surfaces.

2. Increasing the water surface exposed to the air.

3. Lowering the atmospheric pressure.

4. Raising the entering water temperature.

5. Reducing the vapor content of the inlet air.

Performance: The measure of the device's ability to cool water. Usually expressed in terms of cooling a quantity of water from a specified hot water temperature to a specified cold water temperature at a specific wet bulb temperature.

It should be noted and stressed that circulating cooling water device selection is intimately associated with condenser and turbine design and operation.

Though one can make approximate statements on the effect of decreased vacuum, increased pressure, initial temperature, increased reheat temperature, and increased feed water

FIGURE 12
Various Types of Cooling Devices

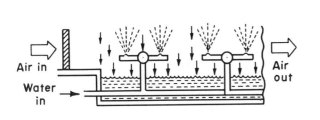

SPRAY POND

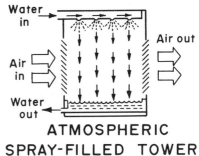

ATMOSPHERIC
SPRAY-FILLED TOWER

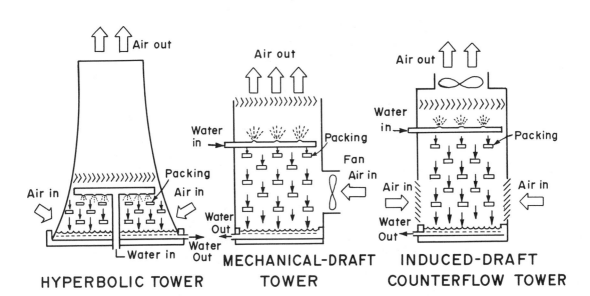

HYPERBOLIC TOWER

MECHANICAL-DRAFT
TOWER

INDUCED-DRAFT
COUNTERFLOW TOWER

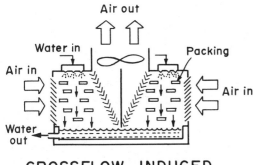

CROSSFLOW INDUCED
DRAFT TOWER

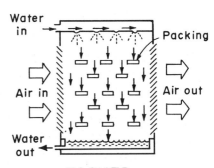

PACKED
ATMOSPHERIC TOWER

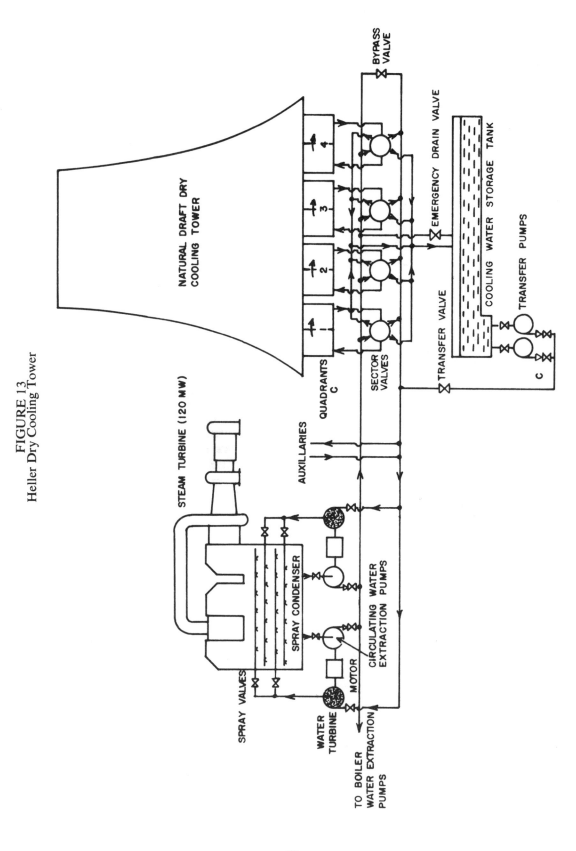

FIGURE 13
Heller Dry Cooling Tower

temperature, the efficiency of the turbine and the resultant heat rate (efficiency—BTU's required to produce 1 kilowatt hour of electricity) for any specific installation for the most economic solution, it is necessary to take into account the interrelationships of the following factors:[326]

1. Average annual duration of temperature and humidity at the site,

2. Condenser heat loads and vacuum correction curves for the turbines desired,

3. Thermal performance of a number of cooling towers,

4. Rough calculation of the pump and pipe network to determine costs and hydraulic characteristics,

5. Performance of a number of condensers designed for the heat loads and cooling ranges under study, and

6. Costs of each component and thereby the amortization charges, fuel costs, plant lifetime load factor, and incremental capability credit.

As LeBailly noted,[327] "There is no practical and simple formula to compare directly the combined performance of different combinations of condensers and cooling towers operating under the same heat load and wet bulb temperature."

Atmospheric Towers

This discussion will be limited to the types of cooling towers that are used with large modern central electric generating stations. It should be noted, however, that in 1967 electric utilities cooling tower requirements were only $20 million out of a total $70 million annual sales of cooling towers.[328] It should be noted, also, that in the time period 1962 to 1967 over 85% of the cooling towers installed in U.S. power generating stations has been of the induced cross-flow design.[329] Atmospheric towers are no longer used because they cannot provide the large cooling capacity required.

Mechanical Draft Wet Cooling Towers

Theory

All wet cooling towers are essentially described by the Merkel formulation (total heat method)[330]

$$\frac{K_G dA}{L c_s} = \frac{dt_i}{i_i - i_G} \qquad (83)$$

where

K_G = mass coefficient of heat transfer

A = surface area

C_s = specific heat of the air/water vapor mixture

L = weight of water circulating per unit time

t_i = interface temperature between water and air

i_i = enthalpy at interface temperature (t_i)

i_G = enthalpy of moist air at temperature t_G

t_G = air temperature

The total heat transfer per unit area per unit time is a function of the difference between the air/water interface and the air temperature, and the difference in concentration of water vapor in the saturated air at air/water interface temperature, and the humidity of the main air stream. The first difference takes into account the sensible heat transfer and the second, the evaporative heat transfer. The two heat transfer forces can be summed into the single equation shown above where the difference in entering and leaving interface water temperature is a function of the differences in enthalpy of the interface air and the moist air. To maximize heat transfer then, the difference in enthalpy must be maximized.

In addition, the heat dissipated can be increased by:

1. Increasing the velocity of the air past the water surface as that increases both the sensible heat transfer and the evaporative cooling;

2. Increasing the surface area of water exposed to the flowing air again increases sensible heat transfer and evaporative heat transfer;

3. Lowering the atmospheric pressure increases the evaporation rate;

4. Raising the entering water temperature will increase the sensible heat loss and evaporation rate; or

5. Reducing the vapor content (humidity) of the inlet air will increase the evaporation rate.

Advantages and Disadvantages of Mechanical Draft Wet Cooling Towers

Mechanical draft towers are divided into two

categories, forced air flow or induced air flow. The advantages and disadvantages of both are listed below.[331]

Forced Draft Towers
Advantages

1. Absolute control over air supply — correct quantities of air and suitable air velocities may be selected.
2. Close control of cold water temperature.
3. Small ground area required.
4. Generally low pumping head.
5. Location of the tower is not restricted.
6. More packing per unit volume of tower.
7. A closer approach and longer cooling range are possible.
8. Capital cost is less than for a natural draft tower.

Disadvantages

1. Horsepower required to operate the fans.
2. Subject to mechanical failure.
3. Subject to recirculation of the hot humid exhaust air vapors into the air intakes.
4. Maintenance costs are high.
5. Operating costs are high.
6. Performance will vary with wind intensity.
7. Not suitable for all climatic conditions.

Induced Draft Towers
Induced draft towers are subdivided into counterflow and crossflow towers.

Counterflow Induced Draft Tower

Advantages

Coldest water contacts the driest air and the warmest water contacts the most humid air. Maximum performance is thus obtained.

Disadvantages

1. Greater resistance to inlet flow of air increases the fan horsepower.
2. Greater resistance to upward flow of air through falling water requires a greater fan horsepower than in cross-flow draft towers.
3. Uneven distribution of air velocities through the filling.
4. Moisture eliminator area restricts air flow.

5. High pumping head necessary.
6. Water load capacity is limited.
7. Hot water distribution system is inaccessible for ready maintenance.
8. High inlet velocities are liable to suck airborne trash and dirt into the plant.

Crossflow Draft Tower

Advantages

1. Low pumping head.
2. Low static pressure drop on the air side.
3. Convenient arrangement of the distribution system.
4. Possible to clean the distribution system while tower in operation.
5. The fill height is approximately equal to the tower height.
6. More air per fan horsepower.
7. More wood fill cooling surface per cubic foot of tower volume.
8. Higher water loadings are possible for a given height.
9. The water temperatures in the basin vary from the center to the edges according to a definite pattern, so that water may be drawn off at a selected temperature.
10. Larger diameter fans can be used so that fewer cells are required for a given capacity.

Disadvantages

1. An insufficient pressure head on the distribution pans to keep the orifices from becoming clogged.
2. Entire water feed exposed to the air which favors growth of algae.
3. A substantial crossflow correction factor needs to be applied to the driving force, particularly where long-range and close approach performances are required. In such cases for some pumping heads a crossflow tower may need more ground area, and more fan horsepower than a counterflow cooling tower.

The major characteristics of induced draft towers — crossflow and counterflow — have been compared.[332]

a. Plan area — counterflow units use less plan area than crossflow. Closer approaches can be obtained by simply increasing the height of the tower if space is a problem.

b. Low draft loss — for a specified applied

71

horsepower a greater air flow is possible in cross-flow towers because of lower static losses and therefore the crossflow tower is more efficient.

c. Thermal performance — crossflow towers can usually attain better thermal performance (i.e., heavier water loadings, longer ranges, and closer approaches) because the air flow distance is independent of fill height and static losses are not increased as length of air travel increases.

d. Low pumping head — static head is constant for crossflow towers but increases as depth of fill increases for counterflow towers. In addition, counterflow units which use spray nozzles require sufficient pressure to break up and distribute the water properly across the tower.

e. Recirculation — recirculation is less likely in crossflow towers than in counterflow towers. When it does occur in crossflow towers only the upper section of the tower is usually affected. In counterflow towers, the recirculation is likely to be more uniformly distributed, hence decreasing efficiency.

f. Multiple cold water temperature withdrawal — in crossflow towers, the temperature of the water in the towers decreases from the edge to the center of the tower. By dividing the tower, it is possible to extract cold water at various temperatures.

g. Winter operation — wide louvers inclined to the vertical enable warm water to be diverted to the sides and to keep the air flow spaces open.

For the above reasons, crossflow towers are most commonly used.

Though forced air flow towers would seem to have overwhelming advantages over induced flow towers such as fewer vibrational problems (fans located at intakes at lower levels), less corrosion since the fans contact the ambient (relatively dry) air, lower noise levels, and slightly more efficiency (since some of the velocity pressure is extracted in useful work), their drawbacks are even greater. Recirculation of the moist, warmed air to the air intakes of the tower is greater than in induced air flow; uniform air distribution through the forced air draft tower is also more difficult to obtain in the large sizes; and fan size is limited in comparison to the size of fan utilized in induced air towers. Conse-quently, for the uses discussed here, induced draft towers are much favored over forced draft towers, and crossflow induced towers over counterflow induced draft towers.

Due to the competition from hyperbolic towers and the ever increasing demand for higher efficiencies many recent improvements have been made in mechanical draft cooling towers.[333] The use of single-flow towers where space is a problem or where the wind flows predominantly in one direction has increased efficiencies more than 10%. The single-flow arrangement can also be useful in preventing freezing by having the northern face solid. Fan and fan cylinder design have been advanced in recent years. Fan blade shape has come closer to the theoretical, and cylinders have incorporated energy recovery shapes. Various towers have been built[334] to maintain their initial efficiency and to handle water temperatures up to 200°F.

Natural-Draft Wet Cooling Towers

Natural-draft towers have been used in Europe as far back as 1907. The first concrete hyperboloidal shell for a natural draft tower was installed in 1916 at the Emma Pit, Heerlen Coal Mine in the Netherlands. Though these towers have been popular in Europe for many years, only recently have any been built in the United States.

Over one half of the generating capacity in the post-World War II construction program for Great Britain[335] utilized natural-draft cooling towers. The first natural-draft tower was built in the United States in late 1962 at the Big Sandy Plant of the Kentucky Power Company. Possibly the reluctance to build natural draft towers in the United States was due to the fact that they appear, in general, to be best suited economically for areas with high humidities. Such areas in this country, eastern and northwestern U.S., also have had abundant water and loose temperature rise restrictions and consequently have used once-through river cooling. Now with greater demands on the available water and tighter water temperature rise restrictions, new attention has been given to the natural-draft towers. Among the advantages of natural-draft towers are long-term maintenance-free operation (no moving parts and a concrete shell), a smaller amount of ground space for multiple towers (no circulation of warm air from one tower to another), reduced piping costs (towers can be located adjacent to plant), no electricity required for operating fans, fewer electrical controls, and less mechanical equipment. Disadvantages include a decreased ability to design as precisely as for mechanical draft towers, and inability to control outlet temperatures as closely as in mechanical draft towers. In addition, because of their large size they tend to dominate the landscape.

McKelvey and Brooke[336] indicate the following advantages and disadvantages for the hyperbolic natural draft towers.

Advantages

1. They produce cooling effects similar to those provided by mechanical draft towers without the mechanical parts and the power required to run them.

2. Maintenance costs are negligible.

3. They can practically never break down.

4. Compared with atmospheric towers they are independent of wind velocity.

5. They can cope with tremendous water loads.

6. They use comparatively small ground areas.

7. The stream of air is in the opposite direction to that of the falling water with the coldest air meeting the colder water first which insures no loss in efficiency.

Disadvantages

1. Resistance to air flow must be kept at a minimum; hence grid sections must be shallow unless some form of film flow is used.

2. The great height necessary to produce the draft.

3. Inlet hot water temperature must be kept hotter than the air dry bulb temperature.

4. Exact control of outlet water temperature is difficult to achieve.

Theory

The basic theory of natural draft cooling towers was recently restated[337] by members of England's Central Electricity Generating Board. They point out that the reason for the hyperbolic shape is aerodynamic and structural rather than thermodynamic. In addition, substantially less material is required for the hyperboloid shape

than for an equivalent straight cylindrical tower.[338] The shell is designed by membrane theory and results in extremely thin sections, as little as 5 inches of reinforced concrete, for towers over 300 feet in height. Details of the analysis including wind measurements, movement of the structure as it becomes saturated, icing, makeup, and integration with the rest of the equipment in the power station are given.

Chilton[339] had pointed out that the Merkel total heat equation was applicable to counterflow cooling towers in which the air flow is known. When the air flow is more variable and is a function of climate and loading, such a direct solution is not possible. Wood and Betts[340] have provided a trial-and-error solution to the problem but the method requires knowledge of constants which are difficult to evaluate. Chilton has derived an approximate solution which is valid if the air becomes saturated during flow through the tower packing. The duty coefficient of the natural draft tower (D), defines the overall capabilities of a tower under all operating conditions. It is essentially a constant over the normal range of operation and is a function of total heat of the air passing through the tower (Δh), BTU/lb dry air, the change of temperature of the water passing through the tower (ΔT, °F), the water load in lb per hour (W_L), and the temperature difference between the dry bulb and wet bulb air temperature (Δt, °F).

$$D = \frac{W_L}{90.59 \, \frac{\Delta h}{\Delta t} \, \sqrt{\Delta t + 0.3124 \, \Delta h}} \tag{84}$$

The draft is due to the difference between the density of the air leaving the tower and the density of the air entering the tower and to the aerodynamic "lift" of the wind passing over the top of the tower.

Cranshaw,[341] in a comprehensive series of tests, has recently studied the variation of performance of natural draft towers and found

1. That the air temperatures should be measured at both top and bottom of the fill.

2. Frictional resistance of the fill and the water to air loading are strongly affected by their change, especially at half loadings where the difference can. be as great as 4 to 5°F.

3. Wind has a lesser effect at full loading than at half loading on tower performance, and

high winds cause a 2°F drop in the cooling range at half load.

4. Performance is negligibly affected by barometric changes.

5. Air stability conditions do reduce the cooling range; an inversion by 1.2°F and a neutral lapse by 0.7°F.

6. Increase in loading, cooling range, and humidity all tend to improve cooling tower performance.

It should be noted that all of these results were obtained under English climatic conditions.

Stern[432] outlines some important items for consideration in choosing natural-draft towers.

1. Temperate climate with above average humidity.

2. Lesser ground area for hyperbolic towers.

3. Height of effluent exhaust will mean less air pollution problems.

Technical factors to be considered are

1. Tower spacing — at least one and one half times base diameter.

2. In crossflow towers, utilizing splash cooling, the fill is outside the shell and increases the diameter of the cold water basin. In the counterflow towers the fill is inside the tower and utilizes film cooling.

3. Packless, or spray filled towers, seem to hold some promise.

Jones[343] has noted that the market for natural draft cooling towers is almost equally divided between crossflow and counterflow towers. For a specific cooled water temperature, higher ambient wet bulb conditions can be tolerated at higher humidities. For a given tower diameter as the approach narrows, the water flow must decrease. The height of towers increased dramatically in the last 10 years, from approximately 300 feet to close to 450 feet. Evaporation losses average about 1.5% of the circulating water, and drift will not exceed 0.2% of the circulating water.

Crossflow towers utilizing wood fill are slightly cheaper to build than counterflow towers. However, the maintenance costs are higher since the wood will eventually have to be replaced, and the operating costs will also be higher, since 10 to 15% higher heads are required than in an equivalent counterflow design.

New Designs

Experimental designs for more efficient cooling towers are under development. Space requirements for both hyperbolic and induced draft towers are so great that combining the best features of both would reduce the space required, improve the efficiency, and reduce the variability of performance. Therefore, mechanical draft has been added to hyperbolic towers to achieve this higher efficiency. The Central Electric Research Laboratories, England,[344] has completed such a design for an assisted-draft tower raising the capacity of a natural-draft tower from 250 to 660 megawatts. In addition, model designs of ellipsoidal form have been tested to separate the airflow and to help create a high draft velocity. Such a slope will help prevent downdrafts into the tower. Such towers would reduce aerodynamic interaction between closely spaced shells (which was partially responsible for the Ferrybridge collapse) and, in addition, would be less conspicuous and occupy less land.[345]

Dry Cooling Towers

The discussion above pertains primarily to wet cooling towers (i.e., those towers where the major mode of heat dissipation is evaporation). For certain cooling conditions such as very high water temperatures, insufficient water, and problems of blowdown disposal, systems that depend primarily upon convection and use air as the transport medium may be preferable. (Dry cooling towers may be either of the natural draft or mechanical draft types). McKelvey and Brooke[346] list the following advantages and disadvantages of a dry cooling tower.

Advantages

1. It can be used where fluids to be cooled are at a high temperature.

2. It eliminates water problems such as availability, chemical treatment, corrosion, spray nuisance, freezing hazard; and fouling.

3. There is no upper limit to which air can be heated.

Disadvantages

1. Normally a dry cooler is less economical than a cooling tower of the ordinary evaporative type.

2. The specific heat of air is only one-fourth that of water.

3. Maintenance costs, e.g., the prevention of corrosion is high.

Theory

Though air cooling has been used for many years for industrial cooling in the petroleum industry and in automobiles, only the recent work of Heller and Fargo[347, 348] has advanced the theory sufficiently to permit larger scale application to central electricity generating stations. As early as 1939, Gesellschaft für Luftkondensation (GEA) had installed an air cooled condenser on a high vacuum stationary steam turbine. The design was the familiar finned tubes with air cooling.[349]

In the unique design of Heller, the exhaust steam from the turbine is sent to a direct contact spray-type jet condenser using cooling water of feed water quality. The condensate is then circulated to the air cooled tower. The entire system is kept under pressure to avoid air leakage.

In addition to a 16 Mw plant in Hungary, two 6 Mw plants in the Soviet Union, two 7.5 Mw plants in Pakistan and an 120 Mw plant in Rugeley, England have been constructed.[350] The Rugeley plant has been in successful operation for over 7 years.[351] The turbine operates at a steam pressure of 1500 psig and steam temperature of 1000°F and reheat to 1000°F and the tower dissipates 587 x 10⁶ BTU/hr with a water savings of 1.8 x 10⁶ gallons per day. The application of the spray condenser to steam turbine was the idea of Heller but the success of the system depended upon having a highly efficient and cheap heat exchanger. This was designed by Fargo. To prevent excessive corrosion 99.5% pure aluminum was used and the joints were mechanical rather than welded or brazed. The tower was the largest in the world at that time reaching 350 feet and having a base diameter of 325 feet. The tower and condenser went into full service in July 1962.

Makeup water in the first three years (for boiler blowdown and other losses) averaged about 2% of boiler requirements. The system has been judged to be a complete success. Designs have been proposed for stations up to 4000 Mw capacity and the English Electric Company expects the dry towers to become a common feature of power stations in the future.[352]

Recently a 150 Mw unit at Ibbenburen, West Germany has been placed in operation and a 200 Mw turbine and two 100 Mw turbines at Gyongyos, Hungary are being installed.[353]

COOLING TOWER PROBLEMS

Though in many respects cooling towers seem to offer overwhelming advantages and to be an absolute necessity in some instances, they are not without their own problems. The wet towers, both mechanical and natural draft, have perhaps more problems than the dry towers but, in general, they are less costly. Some of the problems are described in the following section.

Climatic Effects of Cooling Towers

The use of the atmosphere as a sewer to carry off our wastes is well established, and is favorably looked upon even by professional meteorologists provided that it is done under the proper conditions and with no ill effects.[354] For dry heat injection this feeling is even more pronounced, and professional meteorologists indicate that this discharge into the environment is acceptable and preferred to discharges to the hydrosphere or lithosphere.[355]

The first problem to be answered is the question of major climatic changes due to waste heat rejection to the atmosphere. It has been shown that the amount of heat rejected from our major cities is a not negligible fraction, (3%), of the solar heat reaching the earth's surface for that region.[356] At present over the entire earth's surface the yearly production of man-made energy is about 1/2500 of the radiation balance of the earth's surface. It could equal the surface radiation balance if compounded annually at 10% for 100 years.[357] Its present growth rate is 4% per year. Therefore, even if we alleviate local thermal pollution problems now, the global problem will be upon us in a few decades. In the meantime the following general conclusions have been reached:[358]

(1) "It does appear to be within man's engineering capacity to influence the loss and gain of heat in the atmosphere on a scale that can influence patterns of thermal forcing of atmospheric circulation.

(2) "Purposeful use of this capability is not yet feasible because present understanding of atmospheric and oceanic dynamics and heat exchange is far too imperfect to predict the outcome of such efforts.

(3) "Although it would be theoretically more efficient to act directly on the moving atmosphere, engineering techniques for doing so are not presently available.

(4) "The inadvertent influences of man's activity may eventually lead to catastrophic influences on global climate unless ways can be developed to compensate for undesired effects. Whether the time remaining for bringing this problem under control is a few decades or a century is still an open question.

(5) "The diversity of thermal processes that can be influenced in the atmosphere, and between the atmosphere and ocean, offers promise that, if global climate is adequately understood, it can be influenced for the purpose of either maximizing climatic resources or avoiding unwanted changes."

The second problem is the effect of cooling towers on regional climatology. The professional opinion is that cooling towers will not have a measurable effect on a region's climatology. This opinion is based upon the effect of Lake Meade and similar large-scale man-made systems upon the local climatology.[359]

This was recognized as long ago as 1937 when Holzman[360] stated, "The moisture for precipitation in the United States was derived mainly from the oceans and transported by maritime air masses."

This is not to imply that cities do not have a measurable effect on their climate. Landsberg[361] showed that cities did have a profound influence on climate as illustrated in Table 9 (see above). Landsberg[362] notes that, "The influences of the city on precipitation are not easily unraveled. We can say with reasonable confidence, however, that most of them tend to increase precipitation."

The main causes for increased precipitation are: additional water vapor from combustion processes and factories; added nuclei of condensation and freezing; thermal updrafts from local heating; and updrafts from increased friction turbulence. Chagnon[363] notes the urban effect on precipitation is most pronounced in the colder

TABLE 9

CLIMATIC CHANGES PRODUCED BY CITIES[361]

Element	Comparison with Rural Environs
Contaminants:	
dust particles	10 times more
sulfur dioxide	5 times more
carbon dioxide	10 times more
carbon monoxide	25 times more
Radiation:	
total on horizontal surface	15 to 20% less
ultraviolet, winter	30% less
ultraviolet, summer	5% less
Cloudiness:	
clouds	5 to 10% more
fog, winter	100% more
fog, summer	30% more
Precipitation:	
amounts	5 to 10% more
days with 0.2 in	10% more
Temperature:	
annual mean	1 to 1.5°F more
winter minima	2 to 3°F more
Relative Humidity:	
annual mean	6% less
winter	2% less
summer	8% less
Wind Speed:	
annual mean	20 to 30% less
extreme gusts	10 to 20% less
calms	5 to 20% less

half year of humid continental climates. Though Chagnon[364] notes increases in precipitation over the Champaign-Urbana region, he concludes, "Whether a given city produces, or can produce, a 5, 10, or 15 percent increase in rainfall and number of rainy days in portions of its urban area is a question that neither meteorology nor climatology can at present answer accurately."

More recently Czapski has suggested that the climatic effects "observed" near cities may be due to the effect of the added heat than the condensation and freezing nuclei to which they are usually attributed. Czapski[365] foresees severe consequences of large latent heat and thermal emissions, and notes the following:

(1) Rainfall will be increased downwind for a considerable distance.

(2) Cumulus clouds will prevail most of the time downwind from a large power plant.

(3) Severe thunderstorms and even tornadoes can be caused in very unstable weather by dry and clear heat.

Thom[366] has pointed out that most of the estimates about climatic effects upon cities are speculative and the interpretation of many of the observations about climate is subject to question.

Hall, Russell, and Hamilton[367] have noted that even for major ecological changes such as forests, the most noticeable effect on the microclimate occurs within the forest itself. For shelter belts, the microclimate of the adjacent open areas is most affected.

There is no question that man can modify his local rainfall temporarily as seen in the rain after the fire bombings in Tokyo and Hamburg. It does not seem that the energy and moisture released from cooling towers has yet reached that level, but it is a fruitful area for research.

Fog and Plumes from Wet Cooling Towers

Localized ground fog does occur around many cooling towers when the cooling air water vapor saturation exceeds the capacity of the ambient air for water vapor.[368] It is suggested that the physical location of the cooling tower be adjusted so that the fog does not obscure roadways or other industrial buildings. One method used to reduce ground fog was to increase the height of the discharge stack to enable the fog to diffuse before reaching the ground. In addition, individual discharge stacks were used for each fan cell to reduce the obstruction to air flow and, consequently, reduce downdrafts.[369] Another ingenious method was to heat the cooling tower exhaust air so as to increase its velocity and heat content, and to promote greater turbulence so that the mixtures were below saturation.[370]

The prediction of the size and path of the fog from cooling towers is possible by analogy with smoke plumes. The angle of the spread of the plume usually varies from 18 to 24 degrees and the length of the visible plume was

$$Xp = 5.7 \left(\frac{Vg}{102\,\pi\,Vw} \right)^{\frac{1}{2}} \left(\frac{tge - tgi}{tp - tgi} - 1 \right)^{\frac{1}{2}} \tag{85}$$

where

tg = air or plume temperature, °C
tp = temperature at end of visible plume, °C
Vw = wind speed, ft/s
Vg = total rate from tower, m³/h
    (N.B. All gas volumes refer to 20°C and 1 atm pressure)
Xp = visible plume length, ft

Suffixes

i = inlet
e = exit

Longer and lower plumes occur when air temperature is low, humidity high, and wind speed moderate to high.[371]

Noise

Large, high speed, rotating machinery, and enormous quantities of air moving through restricted spaces cause noise. Consequently, around mechanically induced draft cooling towers, there are going to be high sound levels. A three-step procedure is followed to evaluate noise problems.[372] First, using established values, determine the sound levels that will be acceptable to the neighboring installations. (In most instances, the building served will be insulated from the outside noise.) Background noises may mask the tower noise.

Second, estimate the noise level produced by the cooling tower installation and received at the neighboring site. Make sure that all electrical and water connections are isolated from the tower and the only noise is one from the tower operation itself. Third, if the sound level at the neighboring site is excessive, reduce it by any or a combination of the following methods:

A. Increase the wall density or thickness,

close windows, close air vents, seal cracks, and barrier walls at the site.

B. Change orientation of the tower so that sound is projected in another direction; reflecting walls can be used.

C. Move the tower further away from the critical site.

D. Run fan motors at lower speed at night when load is lowest, but when ambient noise levels may also be lowest.

E. A more expensive but more effective method is to use a lined plenum chamber. A very complete set of curves and tables on acceptable sound levels, noise reduction provided by various walls, background noise, and distance correction values is given in the paper.

In a similar paper, values of noise levels for fans of different horsepower are given.[373] It is also pointed out that fan noise carries further and is more noticeable than water noise. Outlet noise is also greater than inlet noise, and satisfactory noise level reduction can sometimes be obtained by baffling the outlet only.

For more complete details on noise and its control, reference should be made to *Noise Reduction* by Beranek,[374] and to *Handbook of Noise Control* edited by Harris.[375]

Chemical Wastes from Cooling Towers

To promote long serviceable life for cooling towers, algicides and fungicides must be used to prevent damage to the wooden components of cooling towers and corrosion inhibitors to prevent oxidation of the metal parts. These chemicals are a potential pollution hazard if discharged directly to the stream. The major references to these problems are: (1) Primer on Cooling Water Treatment[376] by the National Association of Corrosion Engineers; (2) papers presented at the Cooling Tower Materials and Water Treatment Symposium[377] at the 1965 meeting of the American Chemical Society, Illinois State Water Survey Circular 91; and (3) Berg et al.,[378] Treatise on Water Requirements and Treatments Cost for Cooling Towers. This paper was adopted from the Illinois State Water Survey Circular No. 86.

Scale and foulants adhere to heat transfer surfaces or settle out in low flow areas and reduce the efficiency of the cooling towers. Common types of scales are: calcium carbonate (most common), calcium sulfate, silica, magnesium compounds, iron and manganese compounds, and phosphate compounds. Scale prevention is enhanced by (1) softening the water (but this also increases its corrosiveness); (2) pH adjustment, which is not too satisfactory, since conditions vary widely throughout the cooling tower; (3) chemical treatment of the water by the addition of polyphosphates which slow the rate of precipitation of the scale forming materials; and (4) polyphosphates with organics alone (occasionally for water extremely high in scaling tendencies).

The rate of corrosion can be reduced by the use of anodic and cathodic inhibitors. Typical inorganic, anodic inhibitors are sodium and potassium chromates, phosphates, silicates, nitrites, ferrocyanides, and molybdates. The chromates are most frequently used at high concentrations, more than 200 mg/1 Na_2CrO_4 due to their relatively low cost and superior corrosion inhibition.

Cathodic inhibitors may be salts of zinc, nickel, manganese, and trivalent chromium. They do not appear to be as successful as anodic inhibitors.

Organic inhibitors are sometimes used but their chief use appears to be in conjunction with polyphosphates and chromates to improve their corrosive inhibition.

There is also a wide variety of proprietary compounds which are used to inhibit corrosion. They are mostly based on chromates and polyphosphates with some additives.

Biological fouling not only plugs screens and restricts flow in lines, it can, through the photosynthetic process, release oxygen to accelerate the corrosion process. The most commonly used agent to control biological growth is chlorine in shock loadings. However, the free chlorine residual in the hot return line to the tower should be less than 1 mg/l to reduce chlorine attacks on the tower lumber. For cooling tower sumps where chlorine is not available, a wide variety of toxicants is available.

Cooling tower lumber, to retain its strength, should not be exposed to any strong oxidizing agent and should be protected from biological attack by using wood pressure-treated with chromates and zinc. High temperatures across the wooden surfaces should be avoided to reduce changes in structure and loss of wood material. A vast literature details treatment on these problems by proper material control, operation, and chemical treatment. Among the more useful is that by Dalton[379] with a table of chemicals used for cooling water treatment including purpose, usual dosage, relative effectiveness, waste problems, and relative cost. Dinkel et al.[380] details the advantages of anti-foulants in extending operating periods. High molecular weight polymers prevent deposition by flocculation, and organics such as lignins and tannins disperse the potential fouling materials. Berg et al.[381] showed that chemical costs could be reduced if the mineral concentration in the cycled water could be increased 5 to 10 times rather than the common practice of 1 to 2 times. Recommendations for increasing the mineral concentration had previously been made by Dennan.[382] Berg et al.[383] gave estimates of all the various costs associated with cooling towers and for comparison purposes with once through cooling. These data are used in Cootner and Löf's[384] *Water Demand for Steam Electric Generation*. Details of the proper maintenance of cooling towers is given by Willa.[385] The right materials for use in cooling towers with information on their service under various exposure conditions is presented by Nelson.[386] For normal conditions, cast iron, modular iron, and galvanized steel for major mechanical and support parts; cast aluminum for blades; wrought aluminum for splash bars; copper alloys for nails; and redwood or pressure treated Douglas fir for fill splash bars and framing; Douglas fir plywood for fan decks and hot water basins; glass fiber reinforced polyester for fan blades and cylinders; and polypropylene and high density polyethylene for orifices and spray nozzles are used.

Water effects of cooling towers are due primarily to the chemicals used for corrosion and fouling control. If the blowdown material cannot be rejected directly to the stream, ion exchange resins may be used to remove the chromates on streams and, thereby, may even eliminate blowdown altogether.[387] A report by Hoppe[388] indicates that it may be possible to use a zinc and biodegradable organic without phosphate and chromate and that the blowdown can be discharged to the stream.

Water Quality Effects of Cooling Towers

In passing through wet cooling towers, the temperature is reduced, the water is oxygenated, the BOD reduced, and the chlorine demand lowered.[389] Perhaps the most comprehensive paper on water quality changes in cooling towers has been based upon studies at English power stations ranging from 120 Mw to 600 Mw by Davies.[390] He points out that for a large, modern power station, blowdown can be as much as one million gallons per hour and this can be a sizable fraction of a stream. Since mostly pure water is evaporated, it would be expected that the concentrations of inorganics such as chloride, sodium, sulfate, totally dissolved solids, and conductivity increase directly as a function of the number of recirculations. This was found to be true in most cases for chlorides, in half the cases for sodium, and only in one-third of the cases for sulfate and conductivity. The hydronium ions (pH) at three of the five stations tested were concentrated slightly more than the inorganic constituents which appeared to be due to the loss of carbon dioxide to the air while passing through the tower. Other possible reactions which would lower the concentration of the hydronium ion below the recirculation concentration are oxidation of ammonia to nitric acid, hydrolysis of molecular chlorine to hydrochloric acid, and oxidation of sulfur dioxide. At three of the stations, hardness follows chloride concentrations, but in one station there appears to be precipitation of calcium carbonate.

The main chemical changes to water passing through cooling towers then, are the loss of carbon dioxide, consumption of some of the oxidizable matter, and nitrification of ammonia. These processes are accelerated by the heated waters over their natural rate of occurrence in streams.

Discharge of Condenser Cooling Waters to the Ground

Though few large central station power plants reject their condenser cooling waters to the ground, because of the magnitude of the flows, smaller size units do discharge their effluents in this manner. The Dutch have taken a preliminary look at the problem and concluded that fresh water, for their country, is too valuable and would not be used for cooling purposes and then discharged to the ground.[391] Saline and brackish water can be used, but will eventually raise ground water to temperatures whereby they cannot be used for cooling purposes. The Dutch also conclude,[392] for their country, that the costs of recharging, maintenance, etc., are so great that, if possible, other methods of cooling should be used.

The practice of using ground water for cooling is quite old in this country. By 1937, over 22 million gallons per day were being recharged for water conservation to the ground formations in Kings and Queens Counties in New York[393] at an average temperature rise of 10°F to 15°F. In 1937, a rise of 9°F in the ground water supply due to recharge operations was noted. By 1940, recharge had reached 30 million gallons per day and ground water temperatures had risen as much as 20°F at some of the pumping wells.[394] By 1941, 60,000,000 gallons per day were being recharged and were causing gradual temperature rises in the ground water. These rises, however, were confined to areas near the recharge wells. In Nassau County, Long Island, New York, though high water temperatures have been found near some recharge wells, no widespread thermal pollution had been detected.[395]

It is interesting to note that a recent European paper on the influence of central electric generating stations on the ground water regime near rivers does not even mention thermal pollution as a problem.[396]

A beneficial use of the condenser cooling water discharged to the ground has been noted by Grauby[397] who suggests a modification of production periods by creating a microclimate of several thousand hectares.

Tower Failures

The most spectacular problem with cooling towers was the failure of three natural draft cooling towers at the Ferrybridge C Power Station in England. The prime mode of failure was a vertical tensile failure within the lower part of the structure.[398] The design of the structures was deficient in three areas:

(1) One minute maximum mean wind speeds were used whereas the structure was vulnerable to winds of shorter duration.

(2) No safety margin was left for uncertainties in steady or dynamic wind loadings.

(3) It was not appreciated how much the design could be changed by minor changes in the pressure coefficient distribution. The safety factors were applied to the strength of materials rather than to the load.

No defect or deficiency in the material or workmanship on towers was found which could be considered to have contributed significantly to the collapse. The committee also found that insufficient attention had been paid to the effect of adjacent towers or the building itself in changing the wind loads.[399]

COMPARISON OF COOLING METHODS

Climatic Requirements

Though one frequently sees statements limiting the use of mechanical draft and natural draft towers to certain climatic conditions, these represent, in part, an economic judgment on the part of the writer. For example, McKelvey and Brooke[400] state, "On the other hand, with a mean wet bulb temperature of 40°F there is no economic application at all for mechanically ventilated towers, however low the exhaust heat loading. Canada or countries in North Europe are examples of places where natural draft towers will consequently tend to predominate. With mean wet bulb temperatures of 70°F and over, natural draft towers are not suitable. Countries where such temperatures are habitual may therefore require mechanical equipment."

Fluor Corporation's house organ,[401] states, "Because of this, (air moves in a natural draft tower as a result of a chimney effect created by the difference in density between the moist air inside the shell and the denser air outside) natural draft towers are economically best suited for areas with high humidities, such as western Europe and part of the eastern and northwestern United States."

It is obvious that more than climatic conditions are involved when these statements are made and assumptions of the cost of labor, materials, power, land, etc., are also factored into the economic decision. It should also be noted that the temperature of the returned condensate has ramifications in the choice of the condenser and even the turbine. We restrict ourselves here to the climatic influence on cooling tower choice.

It should be noted that methods of dissipating heat in both natural draft and mechanical draft wet towers are the same, that is, primarily evaporation. Therefore, for wet towers the minimum temperature to which the condensate can be cooled is the wet bulb temperature.

For dry towers, the dry bulb temperature is the minimum temperature to which the condensing water can be cooled.

Cooling Tower Operation

Possibly the best review of current operating experience with mechanical draft cooling towers in the United States is that offered by Waselkow,[402] who details 60 years of experience of the Public Service Company of Colorado. Seventy-six percent of Public Service steam generating capability is on cooling towers. Waselkow also presents data on other mountain and southwestern utility companies' experiences with cooling towers. He cautions about the difficulties in specifying tower performance since it is so intimately related to weather conditions.

The following figures reflect their experience:

(1) One half to three quarters of a gallon of water is consumptively used per kilowatt hour. (For nuclear plant use, 0.9 gallons per kilowatt hour.)

(2) Drift loss is less than 0.1% of the circulating water.

(3) Lake cooling is most trouble free operation, followed by tower operation, and most troublesome is once-through stream flow.

(4) Deterioration of the redwood fill does not follow any consistent pattern.

Natural draft tower experience must be gleaned from European data since only limited American experience is available. Jones[403] estimates evaporative losses at 1½% of the circulating flow and drift at less than 0.2%. Blowdown will vary depending upon water quality but should be less than 2 or 3% of the load even under extreme conditions.

Jones[404] also notes that natural-draft towers, because of their high initial cost, can only be justified where the cooling demand is extremely great and judges that the smallest individual generating unit which will use natural draft towers will be on the order of 500 megawatts.

Dry towers have much smaller losses. At Rugeley,[405] total makeup has been about 1.5% of the boiler feed makeup because of faulty pump glands and cooler leakage. Tower makeup of 1.8×10^6 gallons per day is, of course, not required.[406]

It should be noted that if the existing tower is poorly designed, real estate is at a premium, and repairs are needed on the existing tower, it may be possible to increase its capacity by as much as 60%.[407] Up to 20% increase in capacity can be gained by newer types of fill material which offer greater surface area. The use of positive pressure sprays can increase capacity by 15%. New designs for drift eliminators can give up to 5% better performance. Up to 10% more capacity can be gained by varying air movement through changes in fan blade angle, and speed and size of motor. A properly designed velocity recovery stack can increase capacity as much as 70%. Partitioning of individual fan cells will help to reduce recirculation.

Economic Comparison of Cooling Towers

Can we make an economic comparison of these various alternatives? It is extremely difficult to do so on any objective basis. Jones[408] has pointed out that the physical size and cost of an optimum tower for a given size generating unit can differ in cost by a factor of two because of location, fuel costs, load factors, capability penalties, and capitalization rates. Despite these wide discrepancies, there have been a fairly large number of comparisons made. They are valid only for the conditions assumed and are not generally applicable.

Possibly the most recent and one of the best comparisons was that of Shade and Smith[409] where they compare six different methods of heat rejection for a modern power plant consisting of two 900 Mw units with fossil-fired boilers using present day costs. They emphasize that this is only for general comparison, since the costs for any project are greatly influenced by local conditions. The results are shown in Table 10 (see page 82).

Ritchings and Lotz[410] studied the economics of closed vs. open-cooling water cycles and compared a 200 Mw_t open cycle and 3 closed cycle systems for natural and induced-draft cooling towers, and a mechanical draft fin tube heat exchanger of American design. Results indicate 10% greater investment for closed-type systems

TABLE 10

ECONOMIC COMPARISON OF COOLING METHODS

Heat Rejection Method	Temperature Rise	Inlet Temperature	Condenser Back Pressure (Inches of mercury)	Capacity $/Kw
1. Run of river cooling	17°F	55°F	1.2	5
2. Bay/lake cooling			1.2	6
3. Natural-draft cooling towers—run of river makeup	28°F	70°F	1.5, 2.16	7.5
4. Natural-draft cooling towers—makeup reservoir			1.5, 2.16	11
5. Cooling pond ($1000/acre)		65°F		10
6. Dry cooling towers				22

TABLE 11

ECONOMIC COMPARISON OF COOLING TOWERS

	Induced-Draft Wet Tower	Natural-Draft Wet Tower	Natural-Draft Dry Tower	Induced-Draft Dry Tower	Mechanical Draft Fin Tube Exchanger
Circulating water flow, gpm	91,000	95,000	115,000	160,000	95,000
Cooling range, °F	21	21	17	12	20
Design approach, °F	20	20	38	33	25
Wet bulb temperature, °F	65	62	—	—	—
Dry bulb temperature, °F	—	—	95	95	95
Land area, square feet	18,600	32,000	69,000	22,500	60,000
Total investment cost—millions of dollars	$28,715	$29,580	$31,905	$32,305	$30,980
Back pressure in. Hg (44% station load)	1.3	1.4	1.5	1.9	1.5
Capability loss, $1000	105	69	448	430	292
Annual costs plus capability loss, $1000	6,453	6,476	7,217	7,283	6,979

TABLE 12

EQUIVALENT FUEL COSTS

	River	Lake	Mechanical-Draft Cooling Tower	Natural-Draft Cooling Tower
Capital Costs, $/Kw	1.0	3.5	3.2	7.2
Typical Fuel Costs to give equivalent annual costs ¢/million BTU	25.0	24.4	24.2	23.5

in comparison to an open-cycle wet cooling tower. Annual costs including fixed charges on investment are 10% more. Where water costs are high or where conservation must be practiced, dry towers may be useful.

These were compared against a standard American-made redwood design induced-draft tower. The economic optimum was adopted for each type. These are listed in Table 11 (see page 82). The water requirements, land areas, operating costs, and annual charges are also listed.

Other broad gauged studies of this type by various consulting firms include that of Steur,[411] who for a 150 Mw 1800 psig, 1000/1000F unit at 70% capacity factor, gives in Table 12 (see page 82) comparative capital costs of the various items needed for different type cooling systems.

Weir and Brittain[412] of still another consulting firm, compare the alternatives of various cooling systems. Though they go through many calculations, they basically make a direct comparison only for induced draft towers and run of river cooling. Surprisingly, for the Mississippi River water used as cooling, where the river bed was constantly changing, the cost of the structure was so great as to overshadow the normal savings in run of river water cooling. In that case, for combined total costs, an inland tower to serve two — 160 Mw units was $707,000 cheaper than a run of river plant, and $1,529,000 cheaper than a tower at the river site. Rysselberge[413] compared natural-draft vs. mechanical-draft towers for a 450 Mw plant in the Illinois-Indiana-Ohio area and reached the conclusion that though the first costs of the natural-draft tower were 1.25 times the cost of the mechanical-draft tower, they were more economical when operating costs were taken into account. Rysselberge maintains that on a cost accounting basis only, capital costs of natural-draft towers can exceed mechanically-induced-draft towers by a factor of 1.6 to 1.7 and still be overall competitive.

Smith and Bovier[414] showed that for Keystone Station in Western Pennsylvania, though the cooling tower had an economic penalty of $0.62/Kw for capital and operating costs, the transmission and fuel cost credits for this nine month plant were so overwhelming, $18.54/Kw, that it not only was the choice but that there was no alternative at that location since sufficient cooling water would not otherwise be available.

Ravet[415] compared costs for a 400 Mw unit for natural-draft and induced-draft tower for investor-owned utilities and publicly-owned utilities. The interest charges for the higher initial cost natural-draft tower can make an appreciable difference in the yearly costs of a project. For investor-owned utilities with an 18.8°F approach for a natural-draft tower and 15°F approach for an induced-draft tower, the natural-draft tower is $140,000 lower in cost. For public power utility with an approach of 16.5°F natural draft and 13.5°F induced draft, natural draft has an advantage of $440,000.

It should be noted that the last two economic analyses have been made by Hamon engineers, who have a patented process for constructing hyperbolic natural-draft cooling towers.

Probably the most objective figures are the average prices actually paid for the various types of cooling towers over the past five years. Marley Company,[416] which manufactures both mechanical draft and natural towers, state: "For mechanical induced-draft cooling towers, the average cost per kilowatt of the 12 largest towers purchased in the last 5 years was $1.67/kilowatt. This cost is a delivered and erected cost but does not include the concrete basin, the electrical controls and wiring for same or the piping external to the tower. This cost figure is a weighted average and the spread can be over twice this figure for smaller stations to somewhat less than this figure for the very large stations. Plant design conditions and design ambient temperatures will also affect the cost.

"For eight large natural-draft installations over the last five years, we have an average delivered and erected price of $3.75/kilowatt. This price includes the concrete basin and the cooling tower but does not include piping to the tower. Again, this cost is a weighted average, however; the swing is much narrower."

These average prices do not indicate the costs at a particular site. More important, however, they do not reflect the cost of delivered power since distance to load center and to fuel supply is not represented in the cost figures. Furnish[416] also notes that as of the time of the letter (February 26, 1969) no dry towers had been sold in this country.

TABLE 13

UNIT AND TOTAL COSTS OF COOLING DEVICES

Cooling Device	Area Required (Normalized)	Water Required Gal/Kw/yr	Additional Unit Costs $/Kw[417]	Cost to year 2000 Billions of Dollars[418]
Open Cycle		500,000		
Pond	500-1000			
Mechanical Draft				
1. Wet	1-2	4000-8000	7	11
2. Dry		70	27	
Natural Draft				
1. Wet	1	4000-8000	11	16
2. Dry		70	25	60

Estimates have been made of the cost of the various types of cooling towers[417] and the costs to the year 2,000[418] using conventional wet-induced and natural-draft technology, and extrapolations for dry-cooling towers. The results are shown in Table 13 (see above). It should be pointed out, however, that the cost of cooling towers, while significant and in the aggregate large, places only a small penalty upon the individual user of electricity. It has been estimated that the cost of cooling is about 5 to 7% of total generation costs. The charge for cooling would then be about 0.5 mill per kwhr[420] which would add 24 cents per month to the average household bill in the United States.

Conclusions

As in many environmental problems, the choice of the proper cooling device is not a simple process nor is there a unique result. Because of the relationship between condenser, turbine, and cooling device, a system analysis study must be done to explore a variety of choices. In addition, the other components of the production of power, transmission, fuel supply, water supply, etc., all play major roles in determining the best choice. One must agree with Jones,[419] manager of cooling towers for Hamon-Cottrell, who stated, "I am very much against attempting to make statements which indicate the relative costs of various forms of condenser cooling. I have seen studies resulting in numbers which could be used to indicate any conclusion desired." The figures quoted in this chapter, including the comparison of river cooling, show the validity of the above statement. For planning purposes, the average costs of delivered towers might best be used.

CONCLUSIONS

The best estimates of permissible temperatures and temperature changes for aquatic life and for consumers are given by the National Technical Advisory Committee on Water Quality Criteria.[420] These are succinctly summarized in Table 14[421] (see page 85). It has been shown that new plant sizes will be such that only in a few areas of the country will we be able to meet such standards if run of the river cooling is used. It has also been shown that expectations of increased flow or reduced cooling requirements due to increased thermal efficiencies are not likely to be realized. We, also, cannot expect to find sufficient beneficial use of the degraded heat to substantially diminish the problem. Cooling ponds and towers do offer the technical means to achieve these standards, but at a price.

The choice, then, will be whether to pay the price for preventing the excess heating of our surface waters or to accept the consequences of discharging these heated effluents and let them take their toll. It is, of course, possible to adopt some intermediate role by allowing some increase in the standards for certain stretches of certain rivers and, therefore, require a lesser amount of money for cooling water devices. This choice is not a technical decision but a social choice.

Acknowledgment

The authors wish to acknowledge their debt to Dr. Barry Benedict for his assistance in the preparation of this review.

CONVERSION FACTORS

gallons (U.S.) x 3.785 = liters
BTU x 0.2520 = kilogram calories

psi (pounds per square inch) x 703.1
$\quad$ = kg/m²
inch x 2.540 = centimeters
square miles x 2.590 = square kilometers
square feet x 0.0929 = square meters
pounds x 0.4536 = kilograms
acre x 0.405 = hectares
pounds per square feet x 16.02 =
$\quad$ kilograms per cubic meter
feet x 0.3048 = meters
miles per hour x 1.609 = kilometer per hour
cubic feet per second x 0.02832 =
$\quad$ cubic meters per second

TABLE 14

RÉSUMÉ OF NUMERICAL TEMPERATURE CRITERIA FOR WATER USE[121]

Use	Change from ambient	Upper limit	Rate of change
	°F	°F	°F/hr
Swimming		85[1]	
Water supply	5.0	85	1
Fish and aquatic life:			
Flowing Streams:			
Warm water	5.0	80-93	
Cold water	5.0	68	
Lakes	3.0		
Marine:			
Winter	4.0		1
Summer	1.5		1
Irrigation		85[2]	
Industrial process		95-100[3]	
Cooling		100-120[3]	

[1] Desirable
[2] Also, not less than 55°F
[3] For most uses, temperature is considered acceptable as received

REFERENCES

1. Mathur, S.P., Waste Heat from Steam-Electric Generating Plants Using Fossil Fuels and its Control, FWPCA, Cincinnati, Ohio, 1968,40.

2. Water Resources Council, *The Nation's Water Resources*, U.S. Government Printing Office, Washington, D.C., 1968, 4-3-2.

3. Water Resources Council, *The Nation's Water Resources*, U.S. Government Printing Office, Washington, D.C., 1968, 4-3-5.

4. Water Resources Council, *The Nation's Water Resources*, U.S. Government Printing Office, Washington, D.C., 1968, 4-3-5.

5. Federal Water Pollution Control Administration, Department of the Interior, *The Cost of Clean Water*, II, 127, 1968.

6. Gameson, A.L.H., Hall, H., and Preddy, W.S., *Engineer*, 204, 850, 1957.

7. Johnson, G.A., Personal communication, 1969.

8. Eldridge, E.F., *J. Water Pollut. Contr. Fed.*, 35, 622, 1963.

9. Bregman, J.I., Thermal Pollution Control—Need for Action, paper presented at Thermal Pollution Symposium of the Cooling Tower Institute, New Orleans, Louisiana, 8, 1968.

10. Seaborg, G.T., *Mech. Eng.*, 91, 30, 1969.

11. U.S. Atomic Energy Commission, Nuclear Reactors Built, Being Built, or Planned in the United States as of June 30, 1968.

12. Bregman, J.I., Thermal Pollution Control—Need for Action, paper presented at Thermal Pollution Symposium of the Cooling Tower Institute, New Orleans, Louisiana, p. 3, January 30, 1968.

13. Water Resources Council, *The Nation's Water Resources*, U.S. Government Printing Office, Washington, D.C., 1968, 1.

14. Federal Power Commission, *Hydroelectric Power Evaluation*, U.S. Government Printing Office, Washington, D.C., 1968, 1.

15. Sporn, P., *Nuclear Power Economics, 1962—1967*, U.S. Government Printing Office, Washington, D.C., 1968, 20.

16. Federal Power Commission, *National Power Survey, Part 1*, U.S. Government Printing Office, Washington, D.C., 1964, 67.

17. Rosenthal, M., *A Comparative Evaluation of Advanced Converters—ORNL—3686*, Oak Ridge National Laboratory, Oak Ridge, Tennessee, 1965, 13.

18. Cambel, A.B., *Energy R & D and National Progress*, U.S. Government Printing Office, Washington, D.C., 1964, XXXI.

19. Water Resources Council, *The Nation's Water Resources*, U.S. Government Printing Office, Washington, D.C., 1968, 4-3-3.

20. Water Resources Council, *The Nation's Water Resources*, U.S. Government Printing Office, Washington, D.C., 1968, 4-3-3.

21. Babcock and Wilcox Company, *Steam*, Babcock and Wilcox Company, New York, N.Y., 1963, 10.

22. Babcock and Wilcox Company, *Steam*, Babcock and Wilcox Company, New York, N.Y., 1963, 10.

23. Barak, M., *Chem. Process Eng.*, 95, 1968.

24. Barak, M., *Chem. Process Eng.*, 95, 1968.

25. Archer, D.H., *Mech. Eng.*, 46, 1968.

26. Archer, D.H., *Mech. Eng.*, 46, 1968.

27. Archer, D.H., *Mech. Eng.*, 45, 1968.

28. Barak, M., *Chem. Process Eng.*, 96, 1968.

29. Rosa, R.J. and Hals, F.A., *Industr. Res.*, 1968, 68.

30. Evans, R.K., *Power*, 67, 1968.

31. Gourdine, M.C. and Malcolm, *Mech. Eng.*, 48, 1966.

32. Glaser, *Mech. Eng.*, 24, 1969.

33. National Science Foundation, *Weather Modification*, U.S. Government Printing Office, Washington, D.C., 1966, 123.

34. Federal Water Pollution Control Administration, Department of the Interior, The Cost of Clean Water, II, 127, 1968.

35. McKee, J.E. and Wolf, H.W., *Water Quality Criteria*, State Water Quality Control Board, Sacramento, California, 1963, 283.

36. Streeter, H.W. and Phelps, E.B., *A Study of the Pollution and Natural Purification of the Ohio River*, Public Health Service, U.S. Government Printing Office, Washington, D.C., 1925.

37. O'Connor, D.J. and Dobbins, W.E., *Trans. ASCE*, 123, 650, 1958.

38. Krenkel, P.A. and Orlob, G.T., *Trans. ASCE*, 128, 298, 1963.

39. Gunter, G., *Ecology in Marine Ecology and Paleoecology*, Geol. Society of America, 1, 159, 1957.

40. Streeter, H.W. and Phelps, E.B., Public Health Service Bulletin #146, Washington, D.C., 8, 1925.

41. Gotaas, H.B., *Sewage Works J.*, Vol. 20, 441, 1948.

42. Schroepfer, G.J., *Pollution and Recovery Characteristics of the Mississippi River*, Vol. I, Part III, Report by Sanitary Engineering Division, Department of Civil Engineering, University of Minnesota, (September), 1961.

43. Zanoni, A.E., *Water Research*, 565, 1967.

44. Zanoni, A.E., *J. Water Pollut. Contr. Fed.*, 656, 1969.

45. Butterfield, C.T. and Wattie, E., *Public Health Reports*, 61, 191, 1946.

46. Butterfield, C.T. and Wattie, E., *Public Health Reports*, 59, 1943.

47. Clark, N.A. and Chang, S.L., *J. Amer. Water Works Ass.*, 51, 1310, 1959.

48. Glasstone, S., *A Textbook of Physical Chemistry*, Van Nostrand, New York, 1946, 1087.

49. Fair, G.M. and Geyer, J.C., *Water Supply and Wastewater Disposal*, John Wiley and Sons, New York, 1954, 494.

50. Anon., *Sci. Magazine*, 158, 1967.

51. Krenkel, P.A., Cawley, W.A., and Minch, V.A., *J. Water Pollut. Contr. Fed.*, 37, 1216, 1965.

52. Bohnke, N., *Proc. 22nd Purdue Industrial Waste Conference*, 22, 1961, 761.

53. Bohnke, N., *Proc. Third International Conference on Water Pollution Research*, Pergamon Press, New York, 1966, 169.

54. Drummond, C.E., Jr., Water Pollution Control for the Chattahoochee River, paper presented at Georgia Water Pollution Control Association Conference, Atlanta, 1967, 12.

55. Santala, Veikko, *Heating, Piping, Air Conditioning*, 38, 129, 1966.

56. Elser, H.J., *Progr. Fish. Cult.*, 27, 85, 1961-1962.

57. Iles, R.B., *New Scientist*, 227, 1963.

58. Raymont, J.E.G., *New Scientist*, 1, 10, 1957.

59. Ansell, A.D., *New Scientist*, 14, 1962.

60. Mihursky, J.A., *Bioscience*, 17, 700, 1967.

61. Anon., *Electrical World*, 30, 1968.

62. Anon, *Sci. News*, 93, 169, 1968.

63. Kovaly, H., *Industr. Res.*, 31, 1968.

64. Lerner, W., *Statistical Abstract of the United States—1968*, U.S. Government Printing Office, Washington, D.C., 1968, 657.

65. Iles, R.B., *New Scientist*, 117, 227, 1963.

66. Iles, R.B., *J. Inst. Elec. Eng.*, 9, 1968, 246.

67. Ansell, D., *Ecology*, 44, 397, 1963.

68. Ascione, R., Southwick, W., and Fresco, J.R., *Sci. Magazine*, 153, 1966, 754.

69. Brock, T.D. and Brock, M.L., *Nature*, 209, 1966, 734.

70. Chadwick, W.L., Clark, F.S., and Fox, D.L., *TASME*, 72, 131, 1950.

71. Fox, D.L. and Corcoran, E.F., *Corrosion*, 14, 1958, 131.

72. Renn, E., *J. Amer. Water Works Ass.*, 49, 1957, 410.

73. Fair, G.M. and Geyer, J.C. *Water Supply and Waste Disposal*, John Wiley and Sons, New York, 1954, 658.

74. Velz, C.J., *J. Amer. Water Works Ass.*, 26, 345, 1934.

75. Camp, T.R., Root, D.A., and Bhoota, B.U., *J. Amer. Water Works Ass.*, 32, 1913, 1940.

76. Arnold, G.E., *J. Amer. Water Works Ass.*, 54, 1336, 1962.

77. Crotty, P.A., Feng, T., Skrinde, R.T., and Kuzminski, L.N., First Northeastern Regional Antipollution Conference, University of Rhode Island, 1968.

78. Raney, F., *RICE J.*, 19, 1963.

79. Dingman, S.L., Weeks, W.F., and Yen, Y.C., *Water Resources Research*, 4, 960, 1968.

80. Burke, P.P., Effect of Water Temperature on Discharge and Bed Configuration, Mississippi River at Red River Landing, Louisiana, Corps of Engineers, Vicksburg, Miss., 1966, 3.

81. Burke, P.P., *ASCE Preprint 820*, 12, 1969.

82. Burke, P.P., *ASCE Preprint 820*, 12, 1969.

83. Colby, B.R. and Scott, C.H., *Effects of Water Temperature on the Discharge of Bed Material*, U.S. Government Printing Office, Washington, D.C., 1965, G-1.

84. Colby, B.R., *Fluvial Sediments*, U.S. Government Printing Office, Washington, D.C., 1963, A-24.

85. Colby, B.R., *Fluvial Sediments*, U.S. Government Printing Office, Washington, D.C., 1963, A-24.

86. Franco, J., *Waterways and Harbors Division*, Proceedings of the American Society of Civil Engineers, 94, 344, 1968.

87. Highland, J.T., *Public Works*, 93, 101, 1962.

88. Anderson, E.R., et al., A Review of Evaporation Theory and Development of Instrumentation, U.S. Navy Electronics Laboratory, Report 159, 1960, 1.

89. United States Geological Survey, Water Loss Investigations, Lake Hefner Studies, Prof. Paper 269, 1954, 1.

90. United States Public Health Service, *Water Temperature, Influences, Effects and Control*, Proceedings, 12th Pacific Northwest Symposium on Water Pollution Research, 1963, 1.

91. Edinger, J.E. and Geyer, J.C., *Heat Exchange in the Environment*, Cooling Water Studies for Edison Electric Institute, Johns Hopkins University, 1965, 1.

92. Angström, A., *Geografisca Ann.*, 2, 1920.

93. Schmidt, W., *Annalen der Hydrographic und Maritimen Meteorologic*, 111-124, 169-178, 1915.

94. Anderson, E.R., Water Loss Investigations: Lake Hefner Studies, U.S. Geol. Survey, Prof. Paper 269, 1964, 71.

95. Harbeck, G.E., Koberg, G.E., and Hughes, G.H., The Effect of the Addition of Heat from a Power Plant on the Thermal Structure and Evaporation of Lake Colorado City, Texas, U.S. Geol. Surv., Prof. Paper 272-B, 1959, 1.

96. World Meteorological Organization, Measurement and Estimation of Evaporation and Evapotranspiration, Tech. Note No. 83, Geneva, 1966, 62.

97. Edinger, J.E. and Geyer, J.C., *Heat Exchange in the Environment*, Cooling Water Studies for Edison Electric Institute, Johns Hopkins University, 1965, 18.

98. Bergstrom, R.N., Hydrothermal Effects of Power Stations, paper presented at ASCE Water Resources Conference, Chattanooga, Tenn., May 1968, 6.

99. Anderson, E.R., Water Loss Investigations: Lake Hefner Studies, U.S. Geol. Survey, Prof. Paper 269, 1964, 71-119.

100. Edinger, J.E. and Geyer, J.C., *Heat Exchange in the Environment*, Cooling Water Studies for Edison Electric Institute, Johns Hopkins University, 1965, 17.

101. Edinger, J.E. and Geyer, J.C., *Heat Exchange in the Environment*, Cooling Water Studies for Edison Electric Institute, Johns Hopkins University, 1965, 36.

102. Edinger, J.E. and Geyer, J.C., *Heat Exchange in the Environment*, Cooling Water Studies for Edison Electric Institute, Johns Hopkins University, 1965, 25, 26, 29, 51.

103. Lima, D.O., *Power*, 80, 142, 1936.

104. Throne, R.F., *Power*, 86, 1951.

105. U.S. Geological Survey, Water Loss Investigations, Lake Hefner Studies, Prof. Paper 269, 1954, 1.

106. Anderson, E.R., et al., A Review of Evaporation Theory and Development of Instrumentation, U.S. Navy Electronics Laboratory, Report 159, February 1960, 1.

107. U.S. Geological Survey, Prof. Paper 270, U.S. Government Printing Office, Washington, D.C., 1954, 1.

108. Harbeck, G.E., Jr., et al., Water Loss Investigations: Lake Mead Studies, U.S. Geological Survey, Prof. Paper 298, 1958, 1.

109. Turner, J.F., Jr., Evaporation Study in a Humid Region, Lake Michie North Carolina, U.S. Geological Survey, Prof. Paper 272-G, 1966, 137.

110. Frenkiel, J., *J. Geophys. Res.*, 68, 4991, 1963.

111. Rodgers, G.K. and Anderson, D.V., *J. Fish. Res. Board of Canada*, 18, 617, 1961.

112. Koberg, G.E., Methods to Compute Long-Wave Radiation from the Atmosphere and Reflected Solar Radiation from a Water Surface, U.S. Geological Survey, Prof. Paper 272-F, 1964.

113. Harbeck, G.E., Kobert, G.E., and Hughes, G.H., The Effect of the Addition of Heat from a Power Plant on the Thermal Structure and Evaporation of Lake Colorado City Texas, U.S. Geol. Surv., Prof. Paper 272-B, 1959, 30.

114. Harbeck, G.E., Kobert, G.E., and Hughes, G.H., The Effect of the Addition of Heat from a Power Plant on the Thermal Structure and Evaporation of Lake Colorado City Texas, U.S. Geol. Surv., Prof. Paper 272-B, 1959, 27.

115. Raphael, J.M., *Power Division J.*, American Society of Civil Engineers Proceedings, 1962, 157.

116. Delay, W.H. and Seader, J., Temperature Studies on the Umpqua River, Oregon: Water Temperature, Influences and Effects, Proceedings 12th Pacific Northwest Symposium on Water Pollution Research, Corvallis, Oregon, 1963, 57.

117. Boyer, Peter B., Method of Computing Average Reservoir Temperature; Water Temperature, Influences and Effects, Proceedings 12th Pacific Northwest Symposium on Water Pollution Research, Corvallis, Oregon, 1963, 78.

118. Moore, A.M., Instrumentation for Water Temperature Studies: Water Temperature, Influences and Effects, Proceedings 12th Pacific Northwest Symposium on Water Pollution Research, Corvallis, Oregon, 1963, 101.

119. Orlob, G.T., Prediction of Thermal Energy Distribution in Streams and Reservoirs, Report to California Department of Fish and Game, June 1967, 1.

120. Orlob, G.T. and Selna, Prediction of Thermal Energy Distribution in Deep Reservoirs, Proceedings, 6th Annual Sanitary and Water Resources Engineering Conference, Vanderbilt University, Department of Sanitary and Water Resources Engineering, Technical Report No. 13, 1967, 64.

121. Velz, C.J., Calvert, J.D., Deininger, R.A., Heilman, W.L., and Reynolds, J.Z., Waste Heat Dissipation in Streams, Ponds and Reservoirs with Application to the Duke Power Company Proposed Keowee-Jocassee Developments, Report to U.S. Fish and Wildlife Service, April, 1966, 1.

122. Wunderlich, W.O. and Elder, R.A., The Influence of Reservoir Hydrodynamics on Water Quality, Proceedings, 6th Annual Sanitary & Water Resources Engineering Conference, Vanderbilt University, Dept. of Sanitary & Water Resources Engineering, Technical Report No. 13, 1967, 78.

123. Krenkel, P.A., Parker, F.L., and Thackston, E.L., The Effects of Pumped Storage at Tocks Island on Water Quality, Report to Corps of Engineers, Philadelphia, District, November, 1967, 1.

124. Dake, J.M.K. and Harleman, D.R.F., *Water Resources Research*, 5, 484, 1969.

125. Dutton, J.A. and Bryson, R.A., *Limnology and Oceanography*, 7, 80, 1962.

126. Orlob, G.T., A Mathematical Model of Thermal Stratification in Deep Reservoirs, paper presented at the Annual Meeting of the American Fisheries Society, Portland, Oregon, September, 1965.

127. Bachmann, R.W. and Goldman, C.R., *Limnology and Oceanography*, 10, 1965.

128. Dake, J.M.K. and Harleman, D.R.F., *Water Resources Research*, 5, 486, 1969.

129. Schroepfer, G.J., et al., *Pollution and Recovery Characteristics of the Mississippi River*, Vol I, Part III, Report by Sanitary Engineering Division, Department of Civil Engineering, University of Minnesota, September, 1961.

130. Raphael, J.M., *Power Division J.* of ASCE Proceedings, 88, 159, 1962.

131. Messinger, H., Dissipation of Heat from a Thermally Loaded Stream, Article 104, U.S. Geological Survey Prof. Paper 475-C, C 176, 1963.

132. Garrison, J.M. and Elder, R.A., *International Association for Hydraulic Research*, Leningrad. 1964, 1.

133. LeBosquet, A.M., Jr., *J. N. Eng. Water Works Ass.*, 60, 111, 1946.

134. Gameson, A.L.H. et al., *The Engineer*, 852, 1957.

135. Gameson, A.L.H., Gibbs, S.W., and Barrett, M.J., *Water and Water Eng.*, 63, 16, 1959.

136. Velz, C.J. and Gannon, J.J., *J. Water Pollut. Contr. Fed.*, 32, 392, 1960.

137. Duttweiler, D.W., A Mathematical Model of Stream Temperature, Thesis, Johns Hopkins University, 1963, 1.

138. Edinger, J.E, *Heat Exchange in the Environment*, Johns Hopkins University, 43, 1965.

139. Halleau, G., *La Tribune du CEBEDEAU*, No. 246 and 252, 1964.

140. Goubet, A., Problemes poses Par la Refrigeration Naturelle Des Cours D'eau, Electricite de France, 1966.

141. Goubet, A., Influence Des Centrales Thermiques Sur Les Cours D'eau, Electricite de France, 1965, 1.

142. Mandelbrot, L., The Cooling of Steam Plant Condensers, translated from the *Bulletin due Centre de Recherches et D'Essais de Chatou,* No. 7, 1964, 1 (Le Refroidissement des Condenseurs des Centrales Electriques Thermiques.)

143. Fleuret, I., Computer Techniques for Estimation of the Cooling Capacity of a River, Electricite de France, HC-091-68/No. 34, 1968. (Estimation de la Capacite de Refrigeration D'une Riviere Par Calcul Automatique)

144. Gras, R., Cooling of Classical Thermal or Nuclear Central Stations Power Plants by Transfer Across Water Surfaces and by Rivers, Electricite de France, HC = 091-68/No. 9, 1968. (Refrigeration Des Centrales Thermiques Classiques ou Nuclearies Par Plan D'eau et Rivieres)

145. Johns Hopkins University, Advanced Seminar.

146. Jaske, R.T., An Evaluation of the Use of Selective Discharge from Lake Roosevelt to Cool the Columbia River, Battelle—Northwest Laboratory, Report No. BNWL-208, 1966.

147. Jaske, R.T., *Water Research*, 2, 1968.

148. Dingman, S.L., Weeks, W.F., and Yen, Y.C., *Water Resources Research*, 4, 349, 1968.

149. Roesner, L.A., Mathematical Models for the Net Rate of Heat Transfer through the Air-Water Interface of a Flowing Stream, PhD. Thesis, University of Washington, 1969.

150. Danckwerts, R.V., *Ind. Eng. Chem.*, 1951.

151. Ward, J.C., *J. Sanitary Engineering Division,* ASCE, 89, 1963.

152. Duttweiler, D.W., A Mathematical Model of Stream Temperature, Thesis, Johns Hopkins University, 1963.

153. Roesner, L.A., Mathematical Models for the New Rate of Heat Transfer through the Air-Water Interface of a Flowing Stream, PhD. Thesis, University of Washington, 1969, 25.

154. National Council on Marine Resources and Engineering Development, United States Activities in Spacecraft Oceanography, U.S. Government Printing Office, 1967, 3.

155. Lukens, John E., Remote Sensing of Thermal Pollution, HRB-Singer Inc., 1, 1968.

156. Parker, D.C. and Wolff, M.F., *Int. Sci. Technol.*, 20, 1965.

157. Robinove, Charles J., *Water Resources Bulletin*, 3, 32, 1967.

158. Strandberg, Carl H., Analysis of Thermal Pollution from the Air, Proceedings, 17th Industrial Wastes Conference, at Purdue Univ., Lafayette, Ind., 1962.

159. Harleman, D.R.F., Hall, L.C., and Curtis, T.G., Thermal Diffusion of Condenser Water in a River during Steady and Unsteady Flows (with application to the T.V.A. Browns Ferry Nuclear Power Plant). Report No. 111, Hydrodynamics Laboratory, Massachusetts Institute of Technology, September, 1968, 79.

160. Harleman, D.R.F., Mechanics of Condenser Water Discharge from Thermal Power Plants, paper presented at Symposium on the Engineering Aspects of Thermal Pollution, Vanderbilt University, August, 1968.

161. Federal Water Pollution Control Administration, Industrial Waste Guide on Thermal Pollution, Pacific Northwest Water Laboratory, Corvallis, Oregon, 1968, 112.

162. Harleman, D.R.F. and Elder, R.A., *J. Hydraulics Division*, ASCE, 91, 1965.

163. Elder, R.A., Thermal Density Underflow Design and Experience, paper presented at Seventh Hydraulics Conference, Iowa Institute of Hydraulics Research, Iowa City, Iowa, June 1958.

164. Harleman, D.R.F. and Garrison, J.M., The Effect of Intake Design on Condenser Water Recirculation, M.I.T. Hydrodynamics Laboratory Technical Report No. 56, 1962.

165. Chikwendu, L.N. and Francis, J.R.D., *J. Institution Water Engineers*, 21, 3, 351, 1967.

166. Harleman, D.R.F., Mechanics of Condenser Water Discharge from Thermal Power Plants, paper presented at Symposium on the Engineering Aspects of Thermal Pollution, Vanderbilt University, August, 1968.

167. Bata, G., *J. Hydraulics Division*, Proceedings of the ASCE, 83, 1957.

168. Tennessee Valley Authority, Brown's Ferry Nuclear Plant Preliminary Intake-Outlet Design Studies, Report No. 63-7, Tennessee Valley Authority, Engineering Laboratory, 1966.

169. Harleman, D.R.F., Mechanics of Condenser Water Discharge from Thermal Power Plants, paper presented at Symposium on the Engineering Aspects of Thermal Pollution, Vanderbilt University, August, 1968.

170. Keulegan, G.H., Eleventh Progress Report on Model Laws for Density Currents; Form Characteristics of Arrested Saline Wedges, U.S. Department of Commerce, National Bureau of Standards, Report 5482, 1957.

171. Tennessee Valley Authority, Brown's Ferry Nuclear Plant Preliminary Intake-Outlet Design Studies, Report No. 63-7, Tennessee Valley Authority, Engineering Laboratory, 1966.

172. Chikwendu, L.N. and Francis, J.R.D., *J. Institution Water Engineers*, 21, 3, 351, 1967.

173. Harleman, D.R.F., Mechanics of Condenser Water Discharge from Thermal Power Plants, paper presented at Symposium on the Engineering Aspects of Thermal Pollution, Vanderbilt University, August, 1968.

174. Harleman, D.R.F. and Stolzenbach, R.D., A Model Study of Thermal Stratification Produced by Condenser Water Discharge, M.I.T. Hydrodynamics Laboratory, Report No. 107, October, 1967, 51.

175. Keulegan, G.H., Interfacial Instability and Mixing in Stratified Flows, National Bureau of Standards Research Paper 2040, 1949.

176. Ippen, A.T. and Harleman, D.R.F., Steady-state Characteristics of Subsurface Flow, National Bureau of Standards Circular 521, 79, 1952.

177. Harleman, D.R.F., *Stratified Flow*, Handbook of Fluid Dynamics, Section 26, McGraw Hill, 1961.

178. Harleman, D.R.F. and Stolzenbach, K.D., A Model Study of Proposed Condenser Water Configurations for the Pilgrim Nuclear Power Station at Plymouth, Massachusetts, Report No. 113, Hydrodynamics Laboratory, Massachusetts Institute of Technology, May, 1969, 53.

179. Bata, G., *J. Hydraulics Division*, Proceedings of the ASCE, 83, 1957.

180. Tennessee Valley Authority, Brown's Ferry Nuclear Plant Preliminary Intake-Outlet Design Studies, Report No. 63-7, Tennessee Valley Authority, Engineering Laboratory, 1966.

181. Harleman, D.R.F. and Stolzenbach, R.D., A Model Study of Thermal Stratification Produced by Condenser Water Discharge, M.I.T. Hydrodynamics Laboratory, Report No. 107, October, 1967, 51.

182. Bata, G., *J. Hydraulics Division*, Proceedings of the ASCE, 83, 1957.

183. Tennessee Valley Authority, Brown's Ferry Nuclear Plant Preliminary Intake-Outlet Design Studies, Report No. 63-7, Tennessee Valley Authority, Engineering Laboratory, 1966.

184. Harleman, D.R.F., Mechanics of Condenser Water Discharge from Thermal Power Plants, paper presented at Symposium on the Engineering Aspects of Thermal Pollution, Vanderbilt University, August, 1968.

185. Harleman, D.R.F., Mechanics of Condenser Water Discharge from Thermal Power Plants, paper presented at Symposium on the Engineering Aspects of Thermal Pollution, Vanderbilt University, August, 1968.

186. Harleman, D.R.F., Mechanics of Condenser Water Discharge from Thermal Power Plants, paper presented at Symposium on the Engineering Aspects of Thermal Pollution, Vanderbilt University, August, 1968.

187. Harleman, D.R.F., Mechanics of Condenser Water Discharge from Thermal Power Plants, paper presented at Symposium on the Engineering Aspects of Thermal Pollution, Vanderbilt University, August, 1968.

188. Harleman, D.R.F. and Stolzenbach, R.D., A Model Study of Thermal Stratification Produced by Condenser Water Discharge, M.I.T. Hydrodynamics Laboratory, Report No. 107, October, 1967, 51.

189. Chikwendu, L.N. and Francis, J.R.D., *J. Institution Water Engineers*, 21, 3, 351, 1967.

190. Fan, Loh-Nien and Brooks, H.H., *J. Hydraulics Division*, ASCE, 92, 423, 1966.

191. Maxwell, W.H.C. and Pazwash, H., Basic Study of Jet Flow Patterns Related to Stream and Reservoir Behavior, Research Report No. 10, University of Illinois, Water Resources Center, 63, 1967.

192. Morton, B.R., *J. Fluid Mechanics*, 5, 151, 1959.

193. Abraham, G., *J. Hydraulics Division*, ASCE, 91, 138, 1965.

194. Abraham, G., *J. Hydraulics Division*, ASCE, 86, 1, 1960.

195. Cederwall, K., Jet Diffusion: Review of Model Testing and Comparison with Theory, Report Hydraulics Division, Chalmers Institute of Technology, Goteburg, Sweden, 1967.

196. Frankel, R.J. and Cumming, J.D., *J. Sanitary Engineering Division*, ASCE, 91, 33, 1965.

197. Fan, Loh-Nien, Brooks, N.H., *J. Sanitary Engineering Division, ASCE*, 92, 296, 1966.

198. Fan, Loh-Nien, Turbulent Buoyant Jets into Stratified or Flowing Ambient Fluids, Report Number Kh-R-15, W.M. Keck Laboratory of Hydraulics and Water Resources, California Institute of Technology, 1967.

199. Wiegel, R.L., *Oceanographical Engineering*, Prentice-Hall, Inc., New York, 1964, Chapter 16.

200. Frankel, R.J. and Cumming, J.D., *J. Sanitary Engineering Division*, ASCE, 91, 33, 1965.

201. Rawn, A.M., Bowerman, F.R., and Brooks, N.H., *J. Sanitary Engineering Division*, ASCE, 86, 65, 1960.

202. Hart, W.E., *J. Hydraulics Division*, ASCE, 87, 171, 1961.

203. Fan, Loh-Nien, Turbulent Buoyant Jets into Stratified or Flowing Ambient Fluids, Report Number Kh-R-15, W.M. Keck Laboratory of Hydraulics and Water Resources, California Institute of Technology, 1967.

204. Abraham, G., *J. Hydraulics Division*, ASCE, 88, 195, 1962.

205. Morton, B.R., Taylor, G.I., and Turner, J.S., *Proc. Roy. Soc. London*, A234, 1, 1956.

206. Morton, B.R., *J. Fluid Mechanics*, 5, 151, 1959.

207. Wiegel, R.L., *Oceanographical Engineering*, Prentice-Hall, Inc., New York, 1964, Chapter 16.

208. Fan, Loh-Nien, Turbulent Buoyant Jets into Stratified or Flowing Ambient Fluids, Report Number Kh-R-15, W.M. Keck Laboratory of Hydraulics and Water Resources, California Institute of Technology, 1967.

209. Csanady, G.T., *J. Fluid Mechanics*, #1164, 22, 225, 1965.

210. Bosanquet, C.H., Horn, G., and Thring, M.W., *Proc. Roy. Soc. London*, A263, 340, 1961.

211. Priestley, C.H.B., *Quart. J. Roy. Meteorol. Soc.*, 82, 165, 1956.

212. Fan, Loh-Nien, Turbulent Buoyant Jets into Stratified or Flowing Ambient Fluids, Report Number Kh-R-15, W.M. Keck Laboratory of Hydraulics and Water Resources, California Institute of Technology, 1967.

213. Callaghan, E.E. and Ruggeri, R.S., Investigation of the Penetration of an Air Jet Directed Perpendicularly to an Air Stream. NACA TN1615, June 1948.

214. Vizel, Y.M. and Mostinskii, I.L., *J. Eng. Physics*, 8, 2, 160, February, 1965; The Faraday Press, translation of *Inzhererno-Fizicheski Zhurnal*.

215. Zeller, R.W., Cooling Water Discharge into Lake Monona, University of Wisconsin, May, 1967, 287.

216. Rawn, A.M., Bowerman, F.R., and Brooks, N.H., *J. Sanitary Engineering Division*, ASCE, 86, 65, 1960.

217. Abraham, G., *J. Hydraulics Division*, ASCE, 91, 138, 1965.

218. Abraham, G., *J. Hydraulics Division*, ASCE, 86, 1, 1960.

219. Frankel, R.J. and Cumming, J.D., *J. Sanitary Engineering Division*, ASCE, 91, 33, 1965.

220. Rawn, A.M., Bowerman, F.R., and Brooks, N.H., *J. Sanitary Engineering Division,* ASCE, 86, 65, 1960.

221. Rawn, A.M., Palmer, H.K., *Trans. Amer. Soc. Civil Engineers*, 94, 1036, 1930.

222. Jen, Y., Wiegel, R.L., Morbarek, I., Surface Discharge of Horizontal Warm Water Jet, Institute of Engineering Research Technical Report HEL-3-3, University of California, Berkeley, California, December, 1964, 44.

223. Burdick, John C., III and Krenkel, P.A., Jet Diffusion under Stratified Flow Conditions, Technical Report Number 11, Sanitary and Water Resources Engineering, Vanderbilt University, 1967.

224. Morton, B.R., Taylor, G.I., and Turner, J.S., *Proc. Roy. Soc. London*, A234, 1, 1956.

225. Morton, B.R., On a momentum-mass flux diagram for turbulent jets, plumes, and wakes. *J. Fluid Mechanics*, 10, 1, 101, February 1961.

226. Ackers, P., Modeling of Heated Water Discharges, paper presented at Symposium on Engineering Aspects of Thermal Pollution, Vanderbilt University, August 1968.

227. Silberman, E. and Stefan, H.G., Modeling the Spread of Heated Water over a Lake Surface, Technical Report, St. Anthony Falls Hydraulic Laboratory, University of Minnesota, 1967.

228. Zeller, R.W., Cooling Water Discharge into Lake Monona, University of Wisconsin, May, 1967, 287.

229. Harremoes, P., Tracer Studies on Jet Diffusion and Stratified Dispersion, paper presented at Third International Conference on Water Pollution Research, Munich, Germany, 1966.

230. Ackers, P., Modeling of Heated Water Discharges, paper presented at Symposium on Engineering Aspects of Thermal Pollution, Vanderbilt University, August 1968.

231. Albertson, M.L., Dia, Y.E., Jensen, R.A., and Rouse, H., *Trans. Amer. Soc. Civil Engineers*, 115, 639, 1950.

232. Reynolds, J.Z., Some water quality considerations of pumped storage reservoirs, Ph.D. Thesis, University of Michigan, 1966.

233. Fan, Loh-Nien, Turbulent Buoyant Jets into Stratified or Flowing Ambient Fluids, Report Number Kh-R-15, W.M. Keck Laboratory of Hydraulics and Water Resources, California Institute of Technology, 1967.

234. Rouse, H., Yih, C.S., and Humphreys, H.W., *Tellus*, 4, 201, 1952.

235. Pratte, B.D. and Baines, W.D., *J. Hydraulics Division*, American Society of Civil Engineers, 92, 53, 1967.

236. Keffer, J.F. and Baines, W.D., *J. Fluid Mechanics*, 15, 481, 1963.

237. Fan, Loh-Nien, Turbulent Buoyant Jets into Stratified or Flowing Ambient Fluids, Report Number Kh-R-15, W.M. Keck Laboratory of Hydraulics and Water Resources, California Institute of Technology, 1967.

238. Zeller, R.W., Cooling Water Discharge into Lake Monona, University of Wisconsin, May, 1967, 287.

239. Ellison, T.H. and Turner, J.S., *J. Fluid Mechanics*, 6, 423, 1959.

240. Zeller, R.W., Cooling Water Discharge into Lake Monona, University of Wisconsin, May, 1967, 287.

241. Jen, Y., Wiegel, R.L., Morbarek, I., Surface Discharge of Horizontal Warm Water Jet, Institute of Engineering Research Technical Report HEL-3-3, University of California, Berkeley, California, December, 1964, 44.

242. Wiegel, R.L., Mobarek, I., Jen, Y., Discharge of Warm Water Jet over Sloping Bottom, Institute of Engineering Research Technical Report HEL 3-4, University of California, Berkeley, California, November, 1964, 60.

243. Sawyer, R.A., *J. Fluid Mechanics*, 17, 481, 1963.

244. Keffer, J.R. and Baines, W.D., *J. Fluid Mechanics*, 15, 481, 1963.

245. Bosanquet, C.H., Horn, G., and Thring, M.W., *Proc. Roy. Soc. London*, A263, 340, 1961.

246. Fan, Loh-Nien, Turbulent Buoyant Jets into Stratified or Flowing Ambient Fluids, Report Number Kh-R-15, W.M. Keck Laboratory of Hydraulics and Water Resources, California Institute of Technology, 1967.

247. Zeller, R.W., Cooling Water Discharge into Lake Monona, University of Wisconsin, May, 1967, 287.

248. Cornell University, Engineering Aspects of Thermal Discharges to a Stratified Lake, College of Engineering, Cornell University, February 1969, 82.

249. Ackers, P., Modeling of Heated Water Discharges, paper presented at Symposium on Engineering Aspects of Thermal Pollution, Vanderbilt University, August, 1968.

250. Fan, Loh-Nien, Turbulent Buoyant Jets into Stratified or Flowing Ambient Fluids, Report Number Kh-R-15, W.M. Keck Laboratory of Hydraulics and Water Resources, California Institute of Technology, 1967.

251. Wiegel, R.L., Mobarek, I., Jen, Y., Discharge of Warm Water Jet over Sloping Bottom, Institute of Engineering Research Technical Report HEL 3-4, University of California, Berkeley, California, November, 1964, 60.

252. Fan, Loh-Nien, Turbulent Buoyant Jets into Stratified or Flowing Ambient Fluids, Report Number Kh-R-15, W.M. Keck Laboratory of Hydraulics and Water Resources, California Institute of Technology, 1967.

253. Taylor, G.I., *Proc. Roy. Soc. London*, 223A, 446, 1954.

254. Elder, J.W., *J. Fluid Mechanics*, 5, 4, 1959.

255. Fischer, H.B., *J. Hydraulics Division*, American Society of Civil Engineers, 93, 187, 1967.

256. Fischer, H.B., Longitudinal Dispersion by Velocity Gradients in Open Channel Flow. Technical Memorandum 64-4, W.M. Keck Laboratory of Hydraulics and Water Resources, California Institute of Technology, Pasadena, California, 1964.

257. Fischer, H.B., A note on the one-dimensional dispersion model, *Air Water Pollut.*, 10, 443, July, 1966.

258. Fischer, H.B., Longitudinal Dispersion in Laboratory and Natural Streams, Report No. KH-R-12, W.M. Keck Laboratory of Hydraulic and Water Resources, California Institute of Technology, 1966.

259. Thackston, E.L. and Krenkel, P.A., Longitudinal mixing in natural streams, *J. Sanitary Engineering Division*, American Society of Civil Engineers, 93, SA5, 67, October, 1967.

260. Parker, F.L., *J. Hydraulics Division, Proc. Amer. Soc. Civil Engineers*, 87, 151, 1961.

261. Fischer, H.B., *J. Hydraulics Division*, American Society of Civil Engineers, 93, 187, 1967.

262. Fischer, H.B., Longitudinal Dispersion in Laboratory and Natural Streams, Report No. KH-R-12, W.M. Keck Laboratory of Hydraulic and Water Resources, California Institute of Technology, 1966.

263. Fischer, H.B., *J. Sanitary Engineering Division*, American Society of Civil Engineers, 94, 927, 1968.

264. Fischer, H.B., *Water Resources Research*, 496, 1969.

265. Edinger, J.E. and E.M. Polk, Initial Mixing of Thermal Discharges into a Uniform Current. Technical Report No. 69-2, National Center for Research and Training in the Hydrologic and Hydraulic Aspects of Water Pollution Control, Vanderbilt University, 1969, 46.

266. Keulegan, G.H., Distorted Models in Density Current Phenomena, Fifth Progress Report on Model Laws for Density Currents, National Bureau of Standards, October 10, 1951.

267. Ackers, P., Modeling of Heated Water Discharges, paper presented at Symposium on Engineering Aspects of Thermal Pollution, Vanderbilt University, August, 1968.

268. Barr, D.I.H., *The Engineer*, 216, 885, 1963.

269. Keulegan, G.H., Interface Mixing in Arrested Saline Wedges, Seventh Progress Report on Model Laws for Density Currents, National Bureau of Standards, June 10, 1955.

270. Silberman, E. and Stefan, H.G., Modeling the Spread of Heated Water over a Lake Surface, Technical Report, St. Anthony Falls Hydraulic Laboratory, University of Minnesota, 1967.

271. Hooper, L.J. and Neale, L.C., *Proc. Boston Soc. Civil Engineers*, 45, 356, 1958.

272. Frazer, W., Barr, D.I.H., and Smith, A.A., *Proc. Institution Civil Engineers*, 39, 23, 1968.

273. Barr, D.I.H., *Proc. Institution Civil Engineers*, 10, 305, 1958.

274. Ackers, P., Modeling of Heated Water Discharges, paper presented at Symposium on Engineering Aspects of Thermal Pollution, Vanderbilt University, August, 1968.

275. Keulegan, G.H., Eleventh Progress Report on Model Laws for Density Currents; Form Characteristics of Arrested Saline Wedges, U.S. Department of Commerce, National Bureau of Standards, Report 5482, 1957.

276. Harleman, D.R.F. and Stolzenbach, K.D., A Model Study of Proposed Condenser Water Configurations for the Pilgrim Nuclear Power Station at Plymouth, Massachusetts, Report No. 113, Hydrodynamics Laboratory, Massachusetts Institute of Technology, May, 1969, 53.

277. Barr, D.I.H., *The Engineer*, 215, 345, 1963.

278. Frazer, W., Barr, D.I.H., and Smith, A.A., *Proc. Institution Civil Engineers*, 39, 23, 1968.

279. Harleman, D.R.F. and Stolzenbach, K.D., A Model Study of Proposed Condenser Water Configurations for the Pilgrim Nuclear Power Station at Plymouth, Massachusetts, Report No. 113, Hydrodynamics Laboratory, Massachusetts Institute of Technology, May, 1969, 53.

280. Maggiolo, O.J. and Spitalnik, J., *J. Hydraulic Research*, 5, 189, 1967.

281. Keulegan, G.H., in *Estuary and Coastline Hydrodynamics*, Ippen, A.T., Ed., Chapter 11, McGraw-Hill, 1966.

282. Ackers, P., Modeling of Heated Water Discharges, paper presented at Symposium on Engineering Aspects of Thermal Pollution, Vanderbilt University, August, 1968.

283. Barr, D.I.H., Proceedings Eighth Congress of International Association for Hydraulic Research, Montreal, 6-C, 1959.

284. Barr, D.I.H., *The Engineer*, 215, 345, 1963.

285. Ackers, P., Modeling of Heated Water Discharges, paper presented at Symposium on Engineering Aspects of Thermal Pollution, Vanderbilt University, August, 1968.

286. Frazer, W., Barr, D.I.H., and Smith, A.A., *Proc. Institution Civil Engineers*, 39, 23, 1968.

287. Keulegan, G.H., Eleventh Progress Report on Model Laws for Density Currents; Form Characteristics of Arrested Saline Wedges, U.S. Department of Commerce, National Bureau of Standards, Report 5482, 1957.

288. Barr, D.I.H., *The Engineer*, 215, 345, 1963.

289. Ackers, P., Modeling of Heated Water Discharges, paper presented at Symposium on Engineering Aspects of Thermal Pollution, Vanderbilt University, August, 1968.

290. Ackers, P., Modeling of Heated Water Discharges, paper presented at Symposium on Engineering Aspects of Thermal Pollution, Vanderbilt University, August, 1968.

291. Cotter, T.J. and Lotz, R.W., *J. Power Division—ASCE*, 1961.

292. Hutchinson, G.E., The Thermal Properties of Lakes, *A Treatise on Limnology*, John Wiley & Sons, Inc., New York, 1957, 534.

293. Langhaar, J.W., *Chem. Eng.*, 60, 194, 1953.

294. Velz, C.J. and Gannon, J.J., *J. Water Pollut. Control Fed.*, 32, 392, 1960.

295. Lima, D.O., *Power*, 142, 1936.

296. Mandelbrot, L., Bull. Centre Recherches d'Essais Chatou, 1, 1964.

297. Throne, R.F., *Power*, 86, 1951.

298. Harbeck, G.E., Jr., *The Use of Reservoirs and Lakes for the Dissipation of Heat,* United States Government Printing Office, Washington, D.C., 1953.

299. Harbeck, G.E., Jr., *J. Power Division ASCE*, 90, 1, 1964.

300. Berman, L.D., *Evaporative Cooling of Circulating Water*, Pergamon Press, New York, 1961.

301. McKelvey, K.K. and Brooke, M., *The Industrial Cooling Tower*, Elsevier, Amsterdam, Netherlands, 1959.

302. McKelvey, K.K. and Brooke, M., *The Industrial Cooling Tower*, Elsevier, Amsterdam, Netherlands, 1959.

303. Berman, L.D., *Evaporative Cooling of Circulating Water*, Pergamon Press, New York, 1961.

304. Langhaar, J.W., *Chem. Eng.*, 60, 194, 1953.

305. Cotter, T.J. and Lotz, R.W., *J. Power Division ASCE*, 103, 1961.

306. Velz, C.J. and Gannon, J.J., *J. Water Pollut. Control Fed.*, 32, 399, 1960.

307. Edinger, J.E. and Geyer, J.C., Heat Exchange in the Environment, Edison Electric Institute, New York, New York, 1965, 107.

308. Harbeck, G.E., Jr., The Use of Reservoirs and Lakes for the Dissipation of Hat, United States Government Printing Office, Washington, D.C., 1953, 1.

309. Throne, R.F., *Power*, 89, 1951.

310. Berman, L.D., *Evaporative Cooling of Circulating Water*, Pergamon Press, New York, 1961, 164.

311. Personal communication.

312. Berman, L.D., *Evaporative Cooling of Circulating Water*, Pergamon Press, New York, 1961.

313. McKelvey, K.K. and Brooke, M., *The Industrial Cooling Tower*, Elsevier, Amsterdam, Netherlands, 1959.

314. Moore, W.E., *Electrical World*, 34, 1968.

315. Federal Water Pollution Control Administration, *The Cost of Clean Water*, Vol. II, Detailed Analysis, United States Government Printing Office, Washington, D.C., 1968, 127.

316. Federal Power Commission, Problems in Disposal of Waste Heat from Steam Electric Plants, Federal Power Commission, Washington, D.C., 1969, 4.

317. McKelvey, K.K. and Brooke, M., *The Industrial Cooling Tower*, Elsevier, Amsterdam, Netherlands, 1959.

318. Berman, L.D., *Evaporative Cooling of Circulating Water*, Pergamon Press, New York, 1961.

319. Cootner, P.H. and Löf, G.O.G., *Water Demand for Steam Electric Generation*, Johns Hopkins Press, Baltimore, 1965.

320. Skrotski, B.G.A. and Vopat, W.A., *Power Station Engineering and Economy*, McGraw-Hill, New York, 1960.

321. Babcock & Wilcox Co., "Steam," Babcock & Wilcox Co., New York, 1963.

322. Marley Company, *Cooling Tower Fundamentals and Application Principles*, Marley Co., Kansas City, Missouri, 1967.

323. Cooling Tower Institute, *Cooling Tower Performance Curves*, Cooling Tower Institute, Houston, Texas, 1967.

324. Elonka, S., *Power*, 51, 1963.

325. Agnon, S. and Chia-Yung, *Heating, Piping, Air Conditioning*, 24, 139, 1952.

326. Devereaux, M.B., *Amer. Power Conf. Proc.*, 28, 464, 1966.

327. LeBailly, A.R., *TASME*, 73, 1021, 1951.

328. Fluor-o-scope, Fluor Corporation, Los Angeles, California, Fall, 1967.

329. Hansen, E.P. and Parker, J.J., *Power Eng.*, 38, 1967.

330. Merkel, F., *Zeitschrift des Vereines Deutscher Ingenieure*, 70, 123, 1926.

331. McKelvey, K.K. and Brooke, M., *The Industrial Cooling Tower*, Elsevier, Amsterdam, Netherlands, 1959, 56.

332. Marley Company, *Cooling Tower Fundamentals and Application Principles*, Marley Co., Kansas City, Mo., 1967, 10.

333. Dickey, J.B., Jr., *Power Eng.*, 30, 1964.

334. Ceramic Cooling Tower Company, *Ceramic Cooling Towers*, Ceramic Cooling Tower Company, Ft. Worth, Texas, 1969, 3.

335. Chilton, H., *Proc. I.E.E.*, 99, Part II, 440, 1952.

336. McKelvey, K.K. and Brooke, M., *The Industrial Cooling Tower*, Elsevier, Amsterdam, Netherlands, 1959, 54.

337. Rish, R.F. and Steel, T.F., *J. Power Division, ASCE*, 89, 1959.

338. Jones, W.J., Natural Draft Cooling Towers, presented at Cooling Tower Institute Annual Meeting, 2, Jan. 1968.

339. Chilton, H., *Proc. I.E.E.*, 99, 440.

340. Wood, B.R. and Betts, P., *The Engineer*, 1950.

341. Crawshaw, C.J., *Proc. Inst. Mech. Engineers*, 178, 997, 1963-64.

342. Stern, W.M., *Power*, 60, 1967.

343. Jones, W.J., Natural Draft Cooling Towers, presented at Cooling Tower Institute Annual Meeting, 3, Jan., , 1968.

344. Stubbs, T., *Electrical World*, 108, 1968.

345. Anon. *Engineering*, 202, 399, 1966.

346. McKelvey, K.K. and Brooke, M., *The Industrial Cooling Tower*, Elsevier, Amsterdam, Netherlands, 1959.

347. Heller, L.I. and Fargo, L., *Fifth World Power Conference*, 5, 3651, 1956.

348. Heller, L.I. and Fargo, L., *Sixth World Power Conference*, 6, 2818, 1962.

349. GEA, Air Cooled Steam Condenser, GEA, Bochum, 1969.

350. Jaszay, T. and Tomcsanyi, G., *Transelektro News*, 9, 29, 1968.

351. Anon., *Power Eng.*, 62, 1965.

352. Christopher, P.J., *English Electric J.*, 20, 15, 1965.

353. Jaszay, T. and Tomcsanyi, G., *Transelektro News*, 9, 32, 1968.

354. Smith, M.E., *Brookhaven Lecture Series*, 24, 1, 1963.

355. Thom, C.S., Personal communication, June 23, 1969.

356. Brooks, N.H., *Man, Water, and Waste*, Calif. Inst. of Technology, Pasadena, Calif., 1967, 110.

357. Fletcher, J.O., *Managing Climatic Resources*, Rand Corp., Santa Monica, Calif., 1969, 3.

358. Fletcher, J.O., *Managing Climatic Resources*, Rand Corp., Santa Monica, Calif., 1969, 15.

359. Holzman, B., Personal communication, June 23, 1969.

360. Holzman, B., *The Hydrologic Cycle in Climate and Man*, U.S. Government Printing Office, Washington, D.C., 1941.

361. Landsberg, H.E., *City Air—Better or Worse*, Public Health Service, Cincinnati, Ohio, 1961, 3.

362. Landsberg, H.E., *City Air—Better or Worse*, Public Health Service, Cincinnati, Ohio, 1961, 8.

363. Chagnon, S.A., Jr., *A Climatological Evaluation of Precipitation Patterns over an Urban Area*, Public Health Service, Cincinnati, Ohio, 1961, 45.

364. Chagnon, S.A., Jr., *A Climatological Evaluation of Precipitation Patterns over an Urban Area*, Public Health Service, Cincnnati, Ohio, 1961, 65.

365. Czapski, U.H., *Possible Effects of Thermal Discharges to the Atmosphere*, New York State Dept. of Health, 1968, 19.

366. Thom, C.S., Personal communication, Sept. 18, 1969.

367. Hall, H., Russell, V.S., and Hamilton, C.D., Trees and Microclimate, *Climatology and Microclimatology*, UNESCO, Paris, 1958, 259.

368. Buss, J.R., *Power*, 72, 1968.

369. Baker, D.R., *Oil Gas J.*, 58, 78, 1960.

370. Hall, W.A., *J. Air Pollut. Contr. Ass.*, 12, 379, 1962.

371. Baker, K.G., *Chem. Process Eng.*, 56, 1967.

372. Miller, L. and Long, F.M., *Heating, Piping, Air Conditioning*, 141, 1962.

373. Seelbach, H., Jr. and Oran, F.M., *Sound*, 2, 38, 1963.

374. Beranek, L.L., Ed., *Noise Reduction*, McGraw-Hill, New York, 1960.

375. Harris, C.M., Ed., *Handbook of Noise Control*, McGraw-Hill, New York, 1957.

376. National Association of Corrosion Engineers, Cooling Water Treating, NACE, Houston, Texas, 1960.

377. Selected Papers on Cooling Water Treatment, Illinois State Water Survey Circular, 91, 1966.

378. Berg, B., Lane, R.W., and Larson, T.E., *J. Amer. Water Works Ass.*, 64, 311, 1964.

379. Dalton, T.F., *Water Works Wastes Eng.*, 53, 1965.

380. Dinkel, C.C., Beecher, J.S., and Corwin, S.H., Anti-foulants and Cooling Water Treatment, Rohm and Haas, Philadelphia, 1967.

381. Berg, B., Lane, R.W., and Larson, T.E., *J. Amer. Water Works Ass.*, 64, 311, 1964.

382. Dennon, W.L., *Industr. Eng. Chem.*, 53, 817, 1961.

383. Berg, B., Lane, R.W., and Larson, T.E., *J. Amer. Water Works Ass.*, 64, 311, 1964.

384. Cootner, P.H. and Löf, G.O.G., *Water Demand for Steam Electric Generation*, Johns Hopkins Press, 1965.

385. Willa, J., *Heating, Piping, Air Conditioning*, 109, 1966.

386. Nelson, J.A., *Heating, Piping, Air Conditioning*, 109, 1966.

387. Anon., *Chem. Engin. Use*, 43, 1964.

388. Hoppe, T.C., *Industr. Water Eng.*, 27, 1966.

389. Dalton, T.F., *Water Works Wastes Eng.*, 53, 1965.

390. Davies, I., *J. Air Water Pollut.*, 10, 853, 1966.

391. Dam, J.C. van, *De Ingenieur*, 51, A743, 1968.

392. Bardet, J., *De Ingenieur*, A758, 1968.

393. Leggette, R.M. and Brashears, M.L., Jr., *Trans. Amer. Geophysical Soc.*, 413, 1938.

394. Brashears, M.L., Jr., *Economic Geol.*, 811, 1941.

395. De Luca, F.A., Hoffman, J.F., and Lubice, E.R., Chloride Concentration and Temperature of the Waters of Nassau County, Long Island, New York, Water Resources Commission, Albany, N.Y., 1965, 5.

396. Stundl, Karl, *La Tribune du Cebedeau*, 20, 74, 1967.

397. Grauby, A., *Energie Nuclaire*, 8, 193, 1966.

398. Hannah, I.W., I Resume of the Report on the Failures, Natural Draught Cooling Towers—Ferrybridge and After, Proceedings of the Conference held at the Institute of Civil Engineers, 1967, 5.

399. Anon., *Engineering*, 399, 1966.

400. McKelvey, K.K. and Brooke, M., *The Industrial Cooling Tower*, Elsevier, Amsterdam, Netherlands, 1959, 336.

401. *Fluor-o-scope*, Fall, 1961.

402. Waselkow, C., *Design and Operation of Cooling Towers*, Engineering Aspects of Thermal Pollution, Parker, F.L., and Krenkel, P.A., Eds., Vanderbilt University Press, Nashville, 1969.

403. Jones, W.J., Natural Draft Cooling Towers, paper presented at Cooling Tower Institute Meeting, Jan. 29, 1968.

404. Jones, W.J., Personal communication, July 8, 1968.

405. Anon., *Engineering*, 202, 399, 1966.

406. Christopher, P.J., *English Electric J.*, 20, 15, 1965.

407. McCann, J., *Power*, 110, 67, 1966.

408. Jones, W.J., Natural Draft Cooling Towers, paper presented at Cooling Tower Institute Meeting, Jan. 29, 1968.

409. Shade, W.R. and Smith, A., *Economics of Cooling Water Use*, Engineering Aspects of Thermal Pollution, Parker, F.L., and Krenkel, P.A., Eds., Vanderbilt University Press, 1969.

410. Ritchings, F.A. and Lotz, A.W., *Proc. Amer. Power Conf.*, 416, 1963.

411. Steur, W.R., *Electrical World*, 157, 42, 1962.

412. Weir, George E. and Brittain, James F., *Amer. Power Conf.*, XXV, 565, 1962.

413. Rysselberge, J.R. van, *Power Eng.*, 96, 1959.

414. Smith, F. III and Bovier, R.E., *Proc. Amer. Power Conf.*, XXV, 406, 1963.

415. Ravet, L.C., *Power Eng.*, 50, 1963.

416. Furnish, A.G., Personal communication, February 26, 1969.

417. Kolflat, R., Statement at Hearings before the Senate Subcommittee on Air and Water Pollution of the Committee of Public Works, U.S. Senate, 90th Congress, 2nd Session, 1968, 22.

418. Bergstrom, R.N., Hydrothermal Effects of Power Stations, paper presented at American Society of Civil Engineers, Chattanooga, Tenn., 1968.

419. Löf, G.O.G., *Industr. Water Eng.*, 15, 1966.

420. Federal Power Commission, *Typical Electric Bills—1967*, United States Government Printing Office, Washington, D.C., 1968, 22.

421. Jones, W.J., Personal communication, July 8, 1968.

422. National Technical Advisory Committee to the Secretary of the Interior, Water Quality Criteria, Federal Water Pollution Control Administration, 1968.

423. Energy Policy Staff, Office of Science and Technology, Executive Office of the President, Consideration Affecting Steam Power Plant Site Selection, 1968, 40.